T0265424

CAMBRIDGE LIBRARY COLLECTION
Books of enduring scholarly value

Darwin

Two hundred years after his birth and 150 years after the publication of 'On the Origin of Species', Charles Darwin and his theories are still the focus of worldwide attention. This series offers not only works by Darwin, but also the writings of his mentors in Cambridge and elsewhere, and a survey of the impassioned scientific, philosophical and theological debates sparked by his 'dangerous idea'.

Views of the Architecture of the Heavens

John Pringle Nichol (1804–59) was a Scottish polymath whose major interests were economics and astronomy; he did much to popularise the latter by his writings. He became Regius Professor of Astronomy at Glasgow in 1836, and in the following year published Views of the Architecture of the Heavens which was immediately successful. George Eliot wrote in a letter of 1841, 'I have been revelling in Nichol's Architecture of the Heavens and Phenomena of the Solar System, and have been in imagination winging my flight from system to system, and from universe to universe ...'. Nichol was a supporter of the nebular hypothesis – that stars form in massive and dense clouds of molecular hydrogen which are gravitationally unstable, and coalesce to smaller denser clumps, which then collapse and form stars – which in modified form is the model most widely accepted today.

Cambridge University Press has long been a pioneer in the reissuing of out-of-print titles from its own backlist, producing digital reprints of books that are still sought after by scholars and students but could not be reprinted economically using traditional technology. The Cambridge Library Collection extends this activity to a wider range of books which are still of importance to researchers and professionals, either for the source material they contain, or as landmarks in the history of their academic discipline.

Drawing from the world-renowned collections in the Cambridge University Library, and guided by the advice of experts in each subject area, Cambridge University Press is using state-of-the-art scanning machines in its own Printing House to capture the content of each book selected for inclusion. The files are processed to give a consistently clear, crisp image, and the books finished to the high quality standard for which the Press is recognised around the world. The latest print-on-demand technology ensures that the books will remain available indefinitely, and that orders for single or multiple copies can quickly be supplied.

The Cambridge Library Collection will bring back to life books of enduring scholarly value (including out-of-copyright works originally issued by other publishers) across a wide range of disciplines in the humanities and social sciences and in science and technology.

Views of the Architecture of the Heavens

In a Series of Letters to a Lady

JOHN PRINGLE NICHOL

CAMBRIDGE UNIVERSITY PRESS

Cambridge, New York, Melbourne, Madrid, Cape Town, Singapore,
São Paolo, Delhi, Dubai, Tokyo

Published in the United States of America by Cambridge University Press, New York

www.cambridge.org
Information on this title: www.cambridge.org/9781108005265

This edition first published 1837
This digitally printed version 2009

ISBN 978-1-108-00526-5 Paperback

VIEWS

OF THE

ARCHITECTURE

OF THE

HEAVENS.

IN A SERIES OF LETTERS TO A LADY.

By J. P. NICHOL, LL.D., F.R.S.E.,

PROFESSOR OF PRACTICAL ASTRONOMY IN THE UNIVERSITY
OF GLASGOW.

EDINBURGH:

WILLIAM TAIT, 78, PRINCES STREET ;

SIMPKIN, MARSHALL, AND COMPANY, LONDON ;

AND JOHN CUMMING, DUBLIN.

M.DCCC.XXXVII.

TO

MISS ROSS OF ROSSIE

THESE LETTERS

ARE

RESPECTFULLY INSCRIBED.

I HOPE none of my readers will expect in this volume a general treatise upon Astronomy: I have not written systematically even upon that portion of the science whose results I intend to expose. It has been my sole aim to state what recent times have evolved concerning the vastness of the Universe, in language so plain, that whoever wills, may henceforth look at the Heavens not without something of the emotion which their greatness communicates to the accomplished Astronomer: and if, in performing this task, I sometimes abandon viii

the style of strict discussion, substitute illustration for proof, or give speculation free wing,
perhaps the scientific reader who discovers the
offence, will, if he approves of my plan, account
it venial. The Plates are chiefly taken from
graphic representations by the two Herschels;
—in most cases they will amply compensate
for the want of powerful telescopes.

College, Glasgow,
1*st June,* 1837.

CONTENTS.

CONTENTS

PLATES.

PART I.

ON THE FORM OF THE EXISTING UNIVERSE.

LETTER I.

GENERAL CONSIDERATIONS ON THE SYSTEM OF THE UNIVERSE—SHAPE OF OUR FIRMAMENT.

MADAM,

I HAVE been induced to make the brief series of letters addressed to you, thus public, because of a regret which, I believe, is widely felt, that the discoveries made in recent years, throwing most unexpected light upon the constitution—present and remote—of the Stellar Universe, should longer continue comparatively unknown, and concealed amid the varied and massive collections of our Learned Societies. Unfortunately, I am not at present in a condition to bestow on these discoveries a shadow of original interest, so that, in description, my pen

can have only a borrowed liveliness ; but as the
illustrious men who share the glory of having
achieved such acquisitions for mankind, have,
not unequivocally, shown a disinclination to the
humbler task of reducing them into a popular
system, we have only the choice of consenting,
that matter of unusual importance shall remain
unfitted to fulfil the best purpose of truth—
which is to instruct and elevate the general
mind—or to permit the work to be attempted
by some one with pretensions no higher than
my own.

Previous to the commencement of this century the facts and speculations about to engage
us were unknown in science.   Before then, the
planetary orbits seemed to encircle all accessible space ; they had effectively constituted bounds
to systematic enquiry, for astronomers had never
adventured into greater remotenesses, having,
like the people, gazed at the farther heavens
with vague and incurious eye—content to ad-

mire their beauty and confess their mystery. This period, however, was distinguished by the occurrence of two events, which could not exist in combination, without ensuring important results. The TELESCOPE, formerly of limited range, suddenly assumed a capability of sounding the uttermost profundities ; and the man in whose hands it took on this new efficiency, was possessed of a genius to which all opportunities could be intrusted, for it was adequate to the highest. The rise of SIR WILLIAM HERSCHEL marks the first, and still the greatest epoch of the modern astronomy. He was struck for a discoverer in the finest mould :—mingling boldness with a just modesty, a thirsting after large and general views, with a peculiar sensitiveness in regard of *existing* analogies, and a habit of most scrupulous and dutiful obedience to their intimations—he was precisely the man first to quit paths, which through familiarity were common and safe, and to guide us into regions, dim and remote, where the mind must be a lamp to itself, and walk entirely by lights which well from its own internal fountains.

There is one infallible mark of the rise of an original mind. When you see a man in the midst of his contemporaries, not contesting opinions—not quarrelling—but quietly, and without either ostentation or fear, proceeding to resolve by reason, subjects which had hitherto been in possession of " common belief," depend on it that a signal access of knowledge is awaiting us, for the freshest stamp of divinity is upon that man. Herschel's first remarkable paper gave a promise of this description, and abundantly was it soon fulfilled. It seems to have early occurred to him that the notions—still prevalent—concerning the relation of our firmament,* or whole heavens, to the universe, or rather to infinite space, rest on no better foundation than many long discarded conceptions

* Once for all, and to prevent ambiguity, let me state, that in speaking of *our firmament,* I mean not the solar system, but that entire mass of stars, of which, what we see in a clear night, is the nearest portion.

which found easy acceptance in less advanced
epochs of astronomy.   The usual inference
from the aspects of the sky is, I believe, that *our*
skies are infinite, or that stars, as we see them,
stretch through all space; which, critically examined, appears only a repetition of the old fal-
lacy, that what is great to us must be great
absolutely, and to all beings,—that a system
must be infinite, occupying and constituting
creation, merely because *we* cannot descry its
boundaries, or reckon up its magnitude by the dimensions of our narrow abode.   The firmament,
with its countless and glorious orbs, is doubtless
vast,—perhaps inconceivably so; but, calmly
placing the utmost sphere within our possible
sight, beside the idea of what is really infinite,—
or comparing the vision of man with the reach of
an Almighty eye—it flashes instantly upon us
that we neither have nor can have positive
ground for the assertion that our stars are diffused through all existence.   Herschel proceeded

to refute systematically this common delusion, and to unfold the true scheme of the universe.

The subject is very unusual, and exceedingly apt to bewilder and overwhelm; so that we will most safely enter it by aid of illustration: and one occurs to me which exhibits with some precision the progress of our Discoverer's thoughts. —Call up to your mind an Indian of that old America, when civilisation had not yet disturbed the sombre twilight of its forests; suppose him of a tribe whose wanderings had been confined far within the interior of a range of primeval pines,—how natural for his untutored thought to conceive the wood of his nativity infinite, or that space is all occupied with trees! His eye had never lighted upon one external object,— the forms of his infancy were the forms to which his manhood had been alone accustomed; trees had always environed him, and hemmed in his prospect; so that, on being informed by an instructed traveller, of the existence of free and

wide savannahs, he must have seemed to hear
of something unintelligible and against nature,
and have gazed with that very incredulity which
fills our minds at the idea of the great Firmament being limited like a Forest—of *our*
Infinite being comprehended within Form. But
lo !—in his stray wanderings—at a time when
his gods smiled upon him—the Indian arrives
at a mountain, whose summit reaches beyond the
heights of the gigantic pines. He attempts it,
overcomes its precipices, and sees—a new world !
The forest of his dwelling is mighty, and stretches
far; but America is mightier, and numbers of
forests equal to his, luxuriate upon its plains.
Where is *our* mountain, do you ask,—where the
height which can pierce these skies ? Indeed it is
seldom found. Men wander through centuries, in
ancient ignorance, without reaching or scaling
an elevation capable of showing them beyond it;
but in propitious hour, and after long prepara-
tion, genius and industry descry it, and straight-
way the scales fall from our sight. It was the

TELESCOPE which in this case carried us into
outer regions, and revealed their contents—
hitherto unseen by human eye. And most splendid is the perspective.   Divided from our firmament and each other by measureless intervals,
numerous FIRMAMENTS, glorious as ours, float
through immensity, doubtless forming one stupendous system, bound together by fine relationships.   These remarkable masses are located so
deep in space, that to inferior telescopes they seem
like faint streaks or spots of milky light upon the
blue of the sky ; but the instruments which had
just been summoned into being resolve their mystery, and disclose their myriads of stars.   One of
these objects, perhaps the most brilliant in the
heavens, is represented in Plate I. : it is in the
constellation Hercules.   After all, how easy the
belief to *its* indwellers, that a mass thus surpassingly gorgeous is—infinite.   What wonder although the inhabitant of a planet revolving
around one of its central suns, should have mis
taken his own magnificent heavens for the uni-

Plate I.

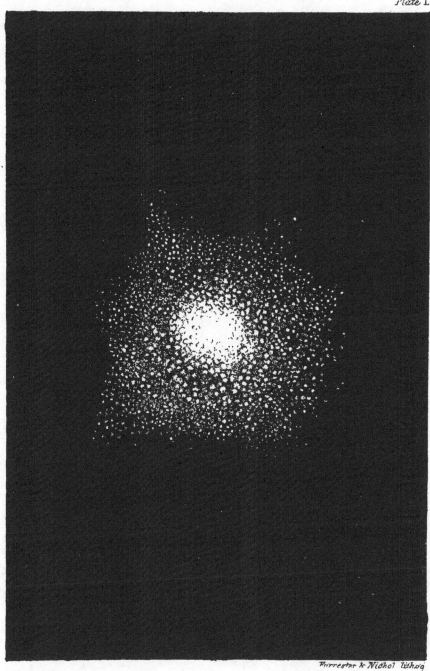

verse, and needed the distant and dim vision of our firmament, appearing to his telescopes as a starry speck, to remove the veil from his mind, and give him juster notions of the majesty of creation !

There are truths which, although startling at first, are found so much in harmony with the scheme of Nature, that we are soon chiefly astonished that they never occurred to us before : and I can conceive circumstances in which the Indian, after the foregoing revelation had been made to him, would not fail to descry among the *internal* aspects of his forest, not only distinct intimations of its limitude, but also of its *peculiar shape*, and even *approximate dimensions*. Think of the appearances, which would be mainly remarked by an observant man, as characteristic of his position, were the forest infinite or very extensive.   In his immediate vicinity the surrounding trees would be well defined, and of the largest proportions ; behind these he would see another range, smaller, but also well defined, and so on through many gradations of size and distinctness, until individual trees could no longer be distinguished, and the view would terminate in an unnamed and vague appearance, which I may be permitted to call a diffused *woodiness*.   But if this peculiar background were not seen in every direction, the light of the sky appearing through the trees in different places, the conclusion would be just and manifest,—*that the forest had not the characteristics of one stretching out indefinitely or even equally on all sides,*—that in some directions its edges were nearer than in others, or that it was merely a *group or stripe of trees having boundaries, and of a particular and ascertainable shape.*   With these fresh lights turn again to the heavens, asking what is the case with them ?   If we were in the interior of an infinite and regular stratum, appearances would necessarily be nearly similar all around us—the aspect of the sky on one side would be almost its picture on every other side. The same, or nearly the same number of visible bodies would, as in the infinite forest, be found every where; and there would come from behind in all directions, through those recesses in which no single star could be descried, something of the same amount of whitish or milky illumination, arising from the combined effulgence of luminaries individually unseen. *But this does not accord with actual phenomena,* which rather agree with the second form of our illustration. It is only when we look towards the MILKY WAY, that these bodies seem to retire indefinitely, and finally to be lost in a diffused *starriness;* and in all other places the intervals between the luminaries are nearly quite dark, *as if there we were closer on the edges of our bed of stars, and therefore saw through it into the external and obscure vacancies of space.* The opinion is thus forced on us anew, that we are in the midst of a mere *group* or *cluster* of stars, and, moreover, that it is a group of peculiar configuration, *narrow, but greatly elongated in the line of the milky way.* Is it not strange that an inference so natural, so palpable, flowing so easily from the phenomena of our brilliant zone, was never drawn before; and that, notwithstanding of that zone shining there, and speaking since " creation's dawn" most invitingly to reason, it should have been left for Herschel, even to so late a day, to rescue it from mythology and mystery, and teach us to look at its countless orbs with emotions loftier than an exciting but useless wonder? But so always has arisen TRUTH—late, late.

Starting with the foregoing general ideas, let us see if we can ascertain, with definiteness, the shape and dimensions of our firmament. In the remaining part of this letter I mean to guide you to the conclusion I have in view, through a series of approximations.

1. Our group, I have said, is certainly an elongated one. A simple diagram will explain how existing appearances are *generally* account-ed for by this hypothesis; and the same mode of illustration points to the first rude modification

which must be imposed on it, to bring it to a more perfect and *minute* consistency with appearances. If a spectator were in a world S in the midst of a stratum or bed of stars, bounded as beneath by the lines C and D, *i. e.* narrow, or, at least, measurable in breadth, or in the

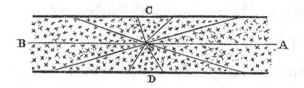

directions, SC, SD, but indefinitely prolonged towards A and B, he would manifestly be engirt by heavens having the general aspect of ours; for, on looking along any line from S towards C or D, he might see through the cluster, and the regions in that quarter would therefore all appear, a comparatively dark ground, bespangled with multitudes of distinguishable luminaries; while in the directions SA and SB— before and behind him—his eye would fail, as ours do when turned to the milky way, in se-

parating the individual bodies, or in recognising
the existence of the remoter masses, otherwise
than in the silvery twilight coming from their
aggregation.   Were our milky zone one regular
belt, we would, it is thus evident, require no
modification of the previous hypothesis ; for the
supposition that the firmament is a regular oblong stratum, would in that case clearly explain
and account for its whole more prominent phenomena ; but the milky way *is not a regular belt.*
An attentive inspection of it, on any favourable
night, will show you, that through one third of
its entire course it *divides into two branches,*
which, after flowing considerably apart—leaving
between them a comparatively dark space—re-
unite, and again form a single stream ; so that
you require a hypothetical figure, which will
explain why, in looking before you to any part
of that region where the stream is divided, your
prospect terminates, not in *one* milky spot, but in
two, separated by a considerable breadth of space
which has no greater number of stars, and not

much more general illumination, than the *sides* of the cluster. Now, this peculiarity will be satisfied, if we suppose the oblong to *divide at one of its extremities ;* for if the sun, S, were in such a cluster as below, a spectator in it would

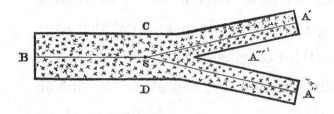

manifestly see one bright spot towards B ; the two equally bright spots, A′ and A″, in the opposite direction; and towards C and D, as before, the background of the heavens would appear comparatively dark, while he would observe a third, but limited dark space, of precisely the same character, towards the vacuity A‴, which separates the divergent branches.

The foregoing figure may be taken with safety as a first, but most general approximation to a section of our vast firmament ; and even as such, I should think it cannot be viewed with an ordinary interest. Perhaps, however, you may

find difficulty in apprehending what our firmament really is, or in inferring its entire shape
from its chart in *section*; I shall, therefore,
on the ground of the foregoing still imperfect considerations, endeavour to construct for
you its complete or solid form.   You know a
common grind-stone? Suppose, first, that over
about one-third of its circumference or rim, the
rim is split in the middle, and along the line of
the rim, which split, however, does not reach so
far down as the centre of the grindstone; also
let the divided parts be somewhat separated towards the middle of the division, running along
as below, and re-entering after their temporary

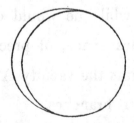

separation.   Suppose, secondly, that the sand-
stone is considerably more porous than stone is,
—then let its minute atoms represent stars, the
pores or intervals, being the interstellar spaces;
and observe what would be the aspect of his

heavens to an inhabitant of a sun or world near
the centre of a cluster of such configuration.
It would be precisely that of our own celestial
vault.   Towards its sides the view would be
comparatively unadorned—dark space looming
from beyond the visible stars ; while, in the direction of the circumference, a countless mass of
small remote stars would, although separately
unseen, illume the sky, forming a splendid
zone, divided like our milky way through part
of its shadowy course.   Is it repulsive to
compare our magnificent universe to an object
like a grindstone ? Alas ! endow it once with
Form, and, whatever its actual dimensions, beside the vastness of the infinity which environs
it, all are shrivelled up, and their majesty disappears ! We cannot now speak of our firmament
but as of a limited object, a speck, one single
individual of an unnumbered throng ; we think
of it, in comparison with creation, only as we
were wont to think of one of its own stars.

2. Hitherto we have used merely the eye—now
we take up the telescope.   Herschel employed

this wonderful instrument to show him the very minute irregularities of our firmament, upon a principle easily understood. Suppose you were somewhere in a crowd—say, in a church filled with people—would you not, on turning and looking around you in different directions, *see a number of persons, proportionate in some sense to your distance from the extremities of the crowd, or the walls of the church?* If you saw a much greater number, for instance, when looking one way, than when looking another way, the inference would surely be apt and ready, that, in the former direction, the mass or multitude of people extended farther than in the latter; and it must be quite conceivable that an arithmetical rule might be found, by which you could *compute* your relative distances in the two cases, from the ends or limits of the assembly. The rule does exist, and is not complex; nevertheless, I am satisfied here if you recognise its *possibility*. Take then a large telescope — one which can go deeply through space—turn it in all directions, count the number of stars in its

field in each position, calculate upon this basis
your corresponding distance from the extremities of the cluster, and you have the means to a
certain extent of accurately charting the firmament. Herschel was fired by this idea. He
termed the process a gaging or sounding the
heavens—casting out lines, as we do at sea, to
fathom and record its profundities. With his
peculiar ardour and perseverance, he accomplished at least 700 known observations with a view
to determine the elements of a suitable and
accurate sketch :—Then supposing S the position of our sun, and drawing in the due directions lines determined in length by the arithmetical rule spoken of, as corresponding to depths
indicated by the quantity of stars in each *gage*,

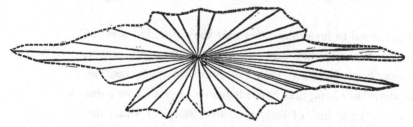

the mere joining of the extremities of these lines
gave him the section required. The above wood-
cut is purely imaginary, being simply intended

to exhibit the rationale of Herschel's *method*;
the result actually obtained by him is shown in
Plate II. It has the aspect, and is larger than the
size under which that firmament—so magnificent
to us—may appear to spectators in remote clusters situated in a line, passing through the milky
way, and nearly over our heads;—spectators who
perhaps are even now marvelling, as they descry
it through their telescopes, what that sprawling
spot may be, which just somewhat, and only in
one trifling point, bedims the azure of their
heavens! *

3. There is obviously one source of error in
Herschel's sketch, which we may be able to
eliminate, and thus approach still nearer the
truth. Throughout the whole speculation you

---

* I proposed to investigate, not only the shape, but also the
dimensions of our cluster. In respect of the latter, we cannot,
of course, hope for more than the remotest approximation.
It was Herschel's idea, that, towards the sides or shallow parts,
there might be a line of forty successive stars, at equal distances from each other, between our sun and its extremities,
while in the direction of the milky way there are in some places
upwards of 900! The milky way, however, is by no means
equally profound. It is, as we will soon see, an extremely
irregular zone.

Plate II.

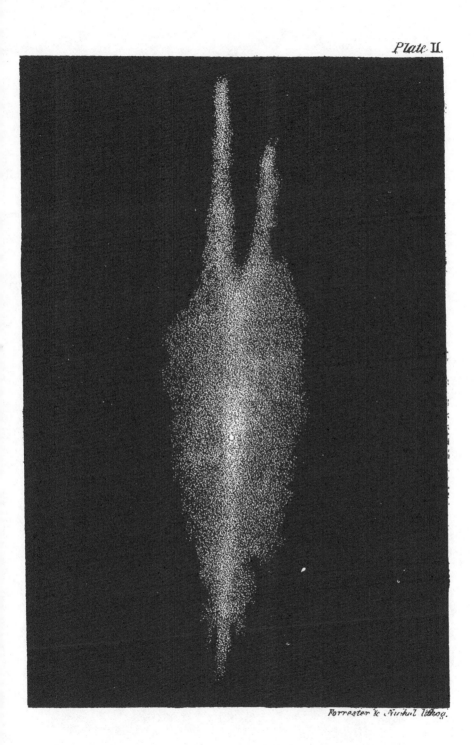

Forrester & Nichol lithog.

find the latent assumption, *that the stars are distributed equably,*—not meaning that there are no
small clusters within the firmament, such as the
Pleiades, or the brilliant Presepe in Cancer, but
that even such clusters are scattered, *generally,*
in all directions, on a principle approaching to indifference.   Our astronomer himself searched in
so far into this matter, and he saw enough to
convince him that his section has only the value
of an approximate representation.   Towards the
milky way, and especially within that zone, the
stars are much more compressed, or closer, than
elsewhere : and I find among Herschel's jour-
nals, notices of various kinds, indicating *breaks*
in the regular progression of stars—absolute
vacuities—chiefly appearing to *detach* the milky
way from the interior mass, and to present it more
as a RING OF STARS, really separated from the rest
of the stratum, but environing it.   Sir John
Herschel decidedly inclines to this opinion :
and he thinks, moreover, that we are not placed
in the centre of the included stratum, but in an
eccentric position, *i. e.* nearer one half of the

ring than its opposite half,—thus accounting for the vastly superior brilliancy of this magnificent girdle in southern latitudes.* The telescope will ultimately resolve all these problems; for, as I will explain in next letter, it not only brings new objects within sight, but also informs us, at the same moment, in what particular zone or sphere they are located; the labour, however, of thus dissecting our massive firmament is scarcely more than begun.—I have now to state to you

* The following are Sir John's words:—" The general aspect of the southern circumpolar region, including in that expression 60° or 70° of S. P. D., is in a high degree rich and magnificent, owing to the superior brilliancy and larger developement of the Milky Way; which, from the constellation of Orion to that of Antinous, is in a blaze of light, strangely interrupted, however, with vacant and almost starless patches, especially in Scorpio, near α Centauri and the cross; while to the North it fades away pale and dim, and is in comparison hardly traceable. I think it is impossible to view this splendid zone, with the astonishingly rich and evenly-distributed fringe of stars of the third and fourth magnitudes, which form a broad skirt to its southern border, like a vast curtain—without an impression, amounting to a conviction, that the Milky Way is not a mere stratum, but an annulus; or, at least, that our system is placed within one of the poorer and almost vacant parts of its general mass, and that eccentrically, so as to be much nearer to the parts about the cross, than to that diametrically opposed to it."

Plate III

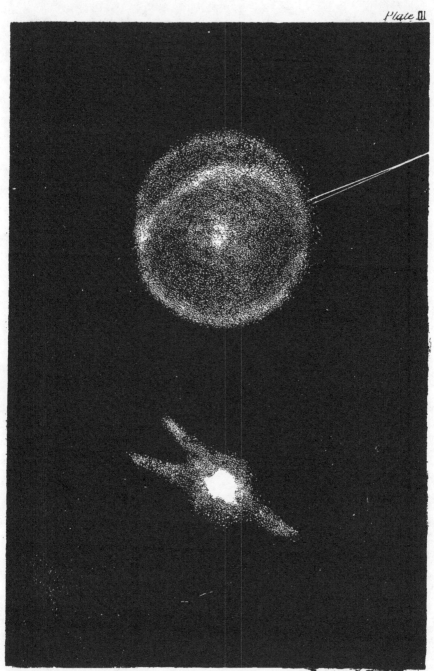

a remarkable circumstance, perhaps the strangest and most unexpected, which modern astronomy has revealed. Although the telescope
has not yet enabled us to lay out the plan of
our cluster, from *interior* surveys, it exhibits
what seems its very *picture*, hung up in external space! Look at Plate III. It represents
an object resting near the outermost range of
telescopic observation, not resolved, but doubtless a great scheme of stars, which is the *fac-
simile* of that to which we belong! It has its
surrounding ring of the precise form we have
been inclined to attribute to our zone; and its
section, figured in the same plate, or the aspect it would take on to a spectator at a vast
distance, looking from the direction of the
white line in the margin, has the closest general resemblance to Herschel's sketch.* Sin-

---

* It adds considerably to the interest attached to this cluster,
that a spectator in it must see ours, precisely as we see it, *i. e.*
sideways; so that our firmament will have, to the inhabitants
of its stars, quite the aspect which it presents to us. The two
figures in Plate III. will explain any ambiguity in pages 17
and 18, where reference is made to the *section* and *solid form*
of our cluster. Perhaps they should now be reperused.

gular affinity of forms! What link, what far-
reaching sympathy can connect these twin
masses—that unfathomed firmament and ours?
What virtue is there in a shape so fantastic that
it should thus be repeated? Or what is the
august law, energising at the opposite extremities of space, which has caused those corresponding shapes to come into being? Prompted
by reverential curiosity, we eagerly put such
questions; but to resolve them baffles our loftiest
philosophies!

Thus then have we entered on new views; we
are introduced to perhaps the grandest phenomena in the Stellar Universe. We have been
raised to an elevation never scaled before, below which Creation and its wonderful arrangements are expanded; and be it never forgotten,
we owe this conquest to the genius of one Man.
But as the circle of knowledge has extended,
the sphere of our recognised ignorance—that
dark sphere which hems in what we know—has
increased likewise. How different is Astronomy

now, taking cognizance of the number of these firmaments, determining their magnitude, and attempting to account for their form, from the Astronomy of a recent day, which was limited to discussions concerning the habits of the small bodies attending our sun, and which, because of its memorable success, boasted that as a science it was complete! There was true prophecy in the exclamation of LA PLACE, who, although then knowing more than any man of the mechanism of the Heavens, said earnestly on his deathbed, " That which we know is little, that which we know not is immense." And the spirit was partaken of by NEWTON in the very flush of his immortal discovery, when, with the modesty of all great minds, beside whose infinite aspirations the highest possible attainment is ever insignificant, he is recorded to have spoken thus :—" I am but as a child standing upon the shore of the vast undiscovered ocean, and playing with a little pebble which the waters have washed to my feet."

LETTER II.

ON THE POWER AND REACH OF TELESCOPES.

It is expedient to digress a little in this place. We have spoken much of the power of telescopes, and of the profundities through which they pierce; and it will be pleasing to you to become satisfied as to the possibility of such power—to be persuaded that I am not amusing you with romance. A brief and simple explanation of the principle on which telescopic power depends, will accomplish this object; and you will then yield me your belief, instead of your mere confidence, when I refer again to what, but a short time ago, would have seemed an incredible fable.

Although not according to the strict mode of expression, it suffices here to say, that we see objects, or that they occasion in a spectator the

Plate IV.

Forrester & Nichol litho.

sensation of seeing, when a certain luminous influence radiating from them, enters the eye in such quantity, as can irritate or excite the nerve of vision. Bright or visible bodies send out this influence in straight lines, and on all sides of them, so that the luminous matter will always be *thinner* the farther it is from the radiating point, as it cannot be dispersed without suffering diffusion—a truth which is pictorially illustrated in Fig. 1, Plate IV., where a luminous point is represented raying out beams. It follows, accordingly, that as the eye is nearer or farther from the source of the rays, the greater or less will be the amount of light received by it; and there will always be a point of distance, intended to be exhibited in the figure, where no more than is barely adequate to cause the sensation of sight, falls upon the retina. This would manifestly be the *remotest limit of visibility* in that case and to that eye; let the spectator retire, although only by the smallest quantity, and

the luminous point would disappear and be lost
in space.

Such being the process of vision, may we hope
ever to extend the boundaries of our sphere, to
reach depths, from which the physical constitution
of humanity might appear to have excluded us ?
There are plainly but two imaginable methods
by which so great an aim may be accomplished,—either we must act upon the shining body
so as to compel it to give forth light of higher
intensity, or we must, if I may say so, *enlarge
the pupil of the eye*, and enable it to receive a
greater number of rays.   The former method is
wholly impracticable—no demonstration is needed to prove it so, for bodies lying profoundly in
space are manifestly remote from our control;
but if by any marvellous artifice we *can produce
a virtual enlargement of the pupil of the eye*, our
success will be equally brilliant and sure.   Nature, indeed, has expressly pointed out this very
plan, for she acts upon it herself.   Noctivagous

creatures have all large pupils, and our own organ of vision enjoys a limited power of expand-
ing in the dusk, so that, when the light is faint,
we take in a larger than usual ray, and discern
objects, which must have been unseen, unless for
that preparatory change.

Turn, without farther preliminary, again to
Plate IV. You know a *lens*—a common burn-
ing-glass? Hold such a glass before the beams
of the sun, and it is seen to compress much of
the light which falls upon its surface into a
bright pencil behind it. Suppose, now, a larger
lens of this description placed, as in Fig. 2, immediately in front of an eye, looking towards a
radiating point which is nearly invisible through
faintness, and notice the change of circumstances.* Instead of the eye receiving, as in
Fig. 1, only a faint and almost imperceptible
quantity of light, it appears taking in *the whole*

---

* It will be recollected that I am not describing the construction of a telescope, which is beyond my province and present intentions: I merely indicate the sources of its power.

mass of rays, which pass through the larger lens in front of it, for the lens has united that quantity, and caused it converge into a pencil sufficiently minute to enter the eye. A portion of the rays which reach the lens from the luminous body, do not indeed pass through it, and are therefore lost to the eye; unless, for which deduction, *the eye in the position referred to would receive as much light, as if its pupil had been enlarged to the dimensions of the lens, and its power would be increased in proportion.* Thus armed, through what a remoteness might we reach that luminary, formerly just escaping us? Without the slightest hazard of its fading away, it might evidently be withdrawn into space, until the whole light compressed by the lens were not more intense at the bright point of its pencil, than the faint ray at the pupil in Fig. 1; and this mighty acquisition has been gained by the simple interposition of a lens! We have been so long accustomed to the telescope, that Wonder has yielded before Familiarity; but perhaps we may yet imagine what amazement spread over Europe when the rumour was first heard, that by a process of such simplicity, the hither boundaries of the unseen had been passed, and remote faintly-twinkling worlds, brought near for the inspection of man! No Arabian tale could have seemed wilder, and no fabled charm of more incredible potency. Power, coming from right use of the laws of the universe, is indeed the true talisman, discovered by the only magic whose gems have perennial sparkle—the exercise of trustful Intellect. Blessed is he who so believes, and so seeks for power—who, through understanding of his own deep nature, and of the system with which the Almighty has environed him, has learned to aim thus wisely at greatness and repose.

Lenses, however, or *refractors*, are not the sole instruments by which light can be collected and compressed. A concave mirror of polished metal, or silvered glass—best of metal—answers the same purpose; only, as seen in the remaining

c

figure of Plate IV., the mirror throws the con-
vergent pencil forward, while refractors throw
it behind them.   I need not remark that the
vital source of power, in either case, is the size
or diameter of the reflecting mirror or refracting
lens; for, according to its diameter, the pupil of
our new eye is great or small.   Refractors of a
moderate size are common, but large ones are very
rare,—Art still failing in the regular construction
of great lenses.   I am assured, however, that instruments exist of this description, through
which a quarter of a mile in the moon is appreciable, and in which one of the mountains of
that enigmatical luminary may be distinctly seen,
and its contour sketched, although magnified so
as to occupy the whole field of view.*   It has

* The magnifying power of a telescope is not the same as
the space-penetrating power, but it depends upon it.   The
space-penetrating power rests wholly in the size of the speculum
or object-glass—it is that power which will recognise a faint
light at a great distance ; the magnifying power again is in the
eye-piece, which is a mere microscope, and enlarges the dimensions of the object brought by the other power into view.   It
is clear, however, that no object which is not presented with

hitherto been easier to work in metal than in glass, and accordingly powerful telescopes, on the reflecting principle, are most frequently employed. Reflectors of nine inches diameter have long been in general use; a considerable number of twice this size have likewise searched the heavens with singular effect; but our *gigantic* instrument is that, perfected by Sir William Herschel, after incredible labour, whose mirror reached the vast magnitude of FOUR FEET diameter. Dwell for a short time on the dimensions and consequent power of this wonderful telescope, and I venture to say you will no longer be sceptical when I speak of seeing into space. If the mirror had reflected all the light which fell upon it, it would virtually have been an eye with a pupil of four feet diameter; that is, it would have been more powerful than the human

great light, can be greatly magnified,—as its light, by being spread or diffused over a large space, would become too thin or faint to be recognised. You may thus infer the penetrating power of the telescope referred to in the text, from the fact that it can permit of so great an amplification of a lunar mountain, and still present it a visible and distinct object.

eye, *by as much as the surface of its enormous disc exceeded the small surface of our pupil!* And making allowance for much light being unavoidably lost, still how great must have been its power! That body is faint indeed, or inconceivably remote, of which it could give us no hope of intimation; and it is no marvel that it sounded our firmament, mighty as it is, and ranged untasked among the abysses of the dark Infinite beyond.   The lustre with which it clothed the bodies in our immediate vicinity is said to have been inexpressibly beautiful.   Herschel himself, intent on far discovery, seldom looked at the larger stars; and, because their blaze injured his eye, he rather avoided their transit.   But he tells us, that at one time, after a considerable sweep with his instrument, " the appearance of Sirius announced itself, at a great distance, like the dawn of the morning, and came on by degrees, increasing in brightness, till this brilliant star at last entered the field of the telescope with all the splendour of the rising sun, and forced me to take my eye from the beautiful sight."

The principles just adverted to, show how we can readily compute the *precise and definite power of our telescopes*, or the distance to which they reach, compared with the naked eye; and such knowledge is of highest importance, as, without it, we could not estimate the relative profundities of the objects which different telescopes reveal. The first element or essential is manifestly the size of the speculum or object-glass, estimated in comparison with the pupil of the eye; and the second is the proportion of light lost in the process of reflection or refraction. If no light were lost, the artificial eye and the natural eye would, as I have said, be efficient, according to the comparative magnitudes of what I may term their respective *pupils;* but as light always is lost,—and the amount may be determined by careful experiment,—a certain deduction must be made from the telescopic power. It were useless to go into the minutiæ of this computation—I merely desire that you comprehend its grounds and rationale. A few *numbers,* however, will be interesting. Herschel considered that his ten-feet telescope had a space-penetrating power of $28\frac{1}{2}$, *i. e.* it could descry a star $28\frac{1}{2}$ times farther off than the naked eye can; to one of his twenty-feet telescopes he assigned the power of 61, and to another of much better construction, the power of 96. The space-penetrating power of the forty-feet instrument he settled at 192! But as you may not have a sufficient idea of the profundities represented by these numbers, I shall convert them into more definite quantities. The depth to which the naked eye can penetrate into space, appears to extend to stars of the twelfth order of distances, *i. e.* it can descry a star twelve times farther away than those luminaries, which, from their superior magnitude, we suppose to be nearest us. Multiply, then, each of the foregoing numbers by twelve, and you have, as a first approximation to the *independent* powers of telescopes, a new series of figures, indicating how much farther they can pierce than the first or nearest range of the fixed stars. In the case of the forty-feet reflector, this number is 2304, which signifies, that if 2304 stars, extended in a straight line beyond Sirius, each separated from the one before it by an interval equal to what separates the still immeasurable Sirius from the earth—the forty-feet telescope would see them all.—I subjoin only one farther statement :—the same instrument could descry a cluster of stars, consisting of 5000 individuals, were it situated three hundred thousand times deeper in space than Sirius probably is; or, to take a more distinct standard of comparison, were it at the remoteness of 11,765,475,948,678,678,679 miles ! *

* The limits to the space-penetrating power of telescopes is manifestly this :—No object fainter than the *general light* of the skies—a light constituted by the intermingling of the rays of all the stars—will ever be seen. Herschel calculated, however, that a telescope, at least three times more powerful than his, might still be used. There are mechanical difficulties in the way of grinding complete metallic specula of such dimension, but might they not be ground in parts? In a letter

It is important to remark, that one telescope
may easily be adjusted to different space-penetrating powers. Its highest power is, of course,
defined and limited by the diameter of the speculum; but we have only to confine that speculum, *to contract its pupil*, by placing an opaque
circular rim of greater or less breadth around

addressed to me by Sir David Brewster, on occasion of our
proposing to erect a new and splendidly-furnished Observatory
in Glasgow, is the following interesting paragraph :—" To such
an Observatory, where the finest achromatic might be accompanied with a better reflecting telescope than has yet been
made, it would be a leading object to delineate with precision
the hills and valleys of the moon.   This planet is much within
our reach ; and an accurate knowledge of the phenomena it
presents, and of the changes these undergo, would be a great and
most interesting contribution to science.   When we compare
the telescope in Newton's time to that of Sir William Herschel's,
we need scarcely despair of discovering the structures erected
by the inhabitants of that luminary.   An achromatic object-
glass of the same size as the speculum of Sir William Herschel's
forty-feet telescope, would certainly accomplish this ; and no
person can say that it is impracticable to do in glass what we
have done in metal.   Had I the means, I would not scruple
to undertake the task of building the lens in zones and seg-
ments."   For the honour of British Science it is to be hoped
that the power of accomplishing what would immortalize his
age, will in some way be afforded to this distinguished philosopher.

the mouth of the tube, to obtain whatever degree of inferior reach we desire. This artifice
enables us to take, without trouble, an observant *walk* through space. We know, for instance, that the naked eye can perceive stars of
the twelfth order of distances, so that whatever
we see without telescopic aid must lie within the
sphere at whose outer circumference stars of that
order are placed. A telescope with a space-penetrating power of 2, or which reaches to the 24th
order of distances, will of course show us much
more than the naked eye; but, whatever additional
it reveals, must, in the main, lie between stars of
the 12th and stars of the 24th order ; so that we
have not only new discoveries, but a view of the
contents of that particular stratum or *layer* of the
firmament, which surrounds our visible sphere.*

* I doubt not you will start a question here :—Is not this
conclusion dependent on the hypothesis, that *stars are all of
the same magnitude ?* For instance, a number of bodies descried by the space-penetrating power referred to, may be *very
small*, and concealed from the naked eye by their *smallness*, not
by their *distance;* so that they may lie much nearer us than

By employing a power of 3, the next layer might
be explored ; and advancing thus, as far as our

even the twelfth order of distances.   I freely grant that if it is
not true that stars are, generally speaking, of much the same magnitude, or, at least, if there is not a magnitude which may be
assumed as the *mean* or *average* one, our speculations will, to
a certain extent, be erroneous, and liable to modification.
Attempts to form a conception of regions quite untrodden must
proceed upon the ground of some assumed principle ; and the
one referred to, is taken, in the present case, as the simplest
and most probable of all.   But there is reason to suppose, that
future investigation will remove doubts even from this position.   In the first place, so soon as we obtain instruments
capable of measuring the actual distances of the stars from
the earth—an achievement perhaps not *much* above present attainment—we will be able to tell exactly what are the relative
sizes of the bodies, within the sphere, whose extreme boundary
has been subjected to measurement ; and as these may be taken
with little admixture of error as representatives of the dimensions of the bodies beyond that sphere, the heavens may be
charted or constructed, on the ground of the information they
will convey.   Secondly, There is an easier and readier mode
of obtaining approximations on this subject.   In last letter I
spoke of *breaks* or *vacuities* in the firmament.   Now, if the stars
are of very various sizes, let us see what is required to constitute such a blank.   Let A

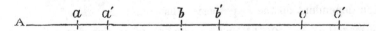

be the place of the telescope, and *a*, *b*, *c*, the places of three orders of stars, descried at the same time, or with the same power
—viz. *a*, the smallest, *b*, the next in order of magnitude, and *c*

powers will sustain us, we may yet complete the speculations referred to near the close of last letter, and take cognizance of the *whole interior structure of the cluster wherein we are.* Mean time, however, we will not proceed with speculation; and therefore I close this letter by quoting, in illustration of the above, Her-

the largest:—it is clear that if stars of the order *c* were wanting between *c* and *c'*; that is, if the space *c c'* were a blank, the telescope would not see it as a blank, *unless the corresponding spaces b b' and a a' were also blanks*—a circumstance so improbable, that we may safely believe it never will occur. The very existence of vacuities thus sustains the likelihood of our hypothesis, that the magnitudes of the stars are not remarkably various; but as the phenomena of these apparently empty spaces, have not been carefully or minutely studied, I am contented to point them out rather as a *means of discovery* in regard of this question. Let me remind you—of what you must never lose sight—that in the spheres we are tracking, we are not on the ground of absolute certainty, nor can we attempt to fix and define all that is to be found there. If we attain a notion of the general forms, or of the *shadows* of mighty truths, our journey will have its success. To carry into such regions the full light of reason, to surmount all difficulties, and clear away all doubts, is reserved for after ages:— it is a welcome thought, that we also, albeit not on earth, may, during these ages, be privileged to gaze with quickened and keener eye over much that is remote from us now.

schel's graphic account of his treatment of one
of the richest and deepest portions of the milky
way. He had previously prepared four telescopes in the manner described—with a series of
gaging or sounding powers, ascending in regular
progression. He adjusted the finder of his seven-
feet telescope to the powers of 2, 3, and 4; to
his night-glass he gave space-penetrating powers
of 5, 6, 7, and 8; to his seven-feet reflector he
gave powers of 9, 10, and upwards, to 17; with
his ten-feet reflector he continued the series from
17 to 28; and, thus amply prepared, he undertook to explore a particular spot. " I selected,"
says he, " the bright spot in the sword handle of
Perseus, as probably a protuberant part of the
milky way in which it is situated. At the time
of observation, not a star in it was visible to the
naked eye. In the finder, with a power of 2, I
saw many stars, and, admitting the eye to reach
to stars at the distance of the twelfth order, we
may conclude that the small stars which were
visible with this low power, are such as to con-

tribute to the brightness of the spot, and that their situation is probably from between the 12th to the 24th order of distances. . . . I then changed the power from 2 to 3, and saw more stars than before, and changing it again from 3 to 4, a still greater number became visible. The situations of these additional stars were consequently between the 24th, 36th, and 48th order of distances. With the gaging power 5 of the night-glass, I saw an increased number of stars; with 6, more stars and whitishness became visible; with 7, more stars, with resolvable whitishness, were seen; and with 8, still more. The stars that now gradually made their appearance, therefore, were probably scattered over the space between the 48th and 96th order of distances. In the seven-feet reflector, with the gaging powers 9 and 10, I saw a great number of new stars; with 11 and 12, a still greater number, and more resolvable whitishness; with 13 and 14, the number of visible stars was increased; and was so again with 15; and with

16 and 17, in addition to the visible stars, there were many too faint to be distinctly perceived." And so did the philosopher go on—invoking space, and summoning up multitudes of worlds—through all the powers of the ten-feet telescope, when I find the following entry : " With the whole space-penetrating power of the instrument, which is 28, the extremely faint stars in the field of view obtained more light, and many still fainter suspected whitish spots could not be verified for want of a still higher gaging power. The stars which filled this field of view were of every various order of telescopic magnitudes ; and, as appeared by these observations, were probably scattered over a space extending from the 204th to the 344th order of distances."

LETTER III.

ASPECTS, FORMS, AND DISTANCES OF REMOTE

FIRMAMENTS.

WE resume our progress. The fact has been already established of the existence of clusters or firmaments, distinct from ours, sustaining an independent position, as individual constituents of creation. Let us now go forth into infinity among these firmaments, and ascertain their character.

The number of such masses is very great. In the northern hemisphere, after making all allowances, those, whose places are fixed, cannot be fewer than between one and two thousand; and you will have a good idea how plentifully they are distributed, by remarking that this is at least equal to the whole number of stars which the naked eye perceives in any ordinary night.

These clusters, the general aspects of which I am now to describe to you, have very various appearances to the telescope. In many of them, individual stars are distinctly defined. As they become more remote, the distances or intervals between the stars diminish, the light also growing fainter; in their faintest stellar aspect, they may be compared to a handful of fine sparkling sand, or, as it is aptly termed, *star-dust;* and beyond this we see no stars, but only a streak or patch of milky light, like the unresolved portions of our own surrounding zone. This is the state in which they are more properly called Nebulæ, and in which there is risk of confounding them with a singular substance not partaking of the nature of stars, but very common in our firmament. It was in reference to this substance, that I spoke of allowance to be made before estimating the number of known clusters. It has not yet been distinguished or separated in many cases from remote and unresolved clusters, a circumstance much to be regretted, the more especially as principles exist, by which an ap-

Plate V

Forrester & Nichol lith.

Plate VI.

proximate determination might, without difficulty, be attained.

In my first letter I drew your attention to
that splendid object in the constellation Hercules, represented in Plate II.   Look at it again,
and imagine its magnificence.   This cluster, in
respect of its leading characteristics, is a good
specimen object, as it is a representative or type
of a very large class.   Notwithstanding the partial irregularity of its outline, it seems almost a
spherical mass, in which, with a degree of greater
compression probably towards the centre, the
stars are pretty equably and regularly diffused,
so that to the inhabitants of worlds near its
central regions, its sky would spangle uniformly
all around, and present no phenomenon like the
milky way in ours.   In Plates V. and VI. are
representations of a few more of these spherical
clusters, some of which, however, show decided
and great compression about the centre—a circumstance which would manifestly much augment the proportionate number of orbs of the

D

first magnitude, in view of those living within
the compressed portion, and thus render their
visible heavens inconceivably brilliant. The
same plates exhibit the degrees of distinctness
with which these clusters appear to us. One of
the figures in Plate VI., you will observe, is a
very faint object, placed near the outermost
verge of the sphere within which our mightiest
instruments can descry individual stars; it is
already in the condition of *star-dust*, almost
fading into an irresolvable nebula.

Firmaments, however, are by no means confined to the spherical form. Our own, and its curious cognate, are exceptions you already recognise, and there are many other equally remarkable shapes. The left hand figure of the upper
line of Plate VI. represents a very singular form
(30 Doradûs), sketched by Mr Dunlop at Paramatta; and we are just given to understand that
it is still stranger than it here appears. Sir John
Herschel, whose return from the Cape of Good
Hope, laden with spoils, all Europe is anxiously

Plate. VII.

Forrester & Nichol. lithog.

expecting, has recently written as follows concerning those Magellanic clouds which have
been long regarded as the wonders of the
Southern skies. " The Nubecula Major and
Minor are very extraordinary objects. The
greater" (see Plate VII. where it is represented
from Dunlop) " is a congeries of clusters of irregular form, globular clusters and nebulæ of
various magnitudes and degrees of condensation,
among which is interspersed a large portion of
irresolvable nebulæ, which may be, and probably is, *star-dust*, but which the power of the
twenty-feet telescope shows only as a general
illumination of the field of view, forming a bright
ground on which the other objects are scattered.
Some of the objects in it are of very singular
and incomprehensible forms; the chief one especially (30 Doradûs), *which consists of a number of loops united in a kind of unclear centre or
knot, like a bunch of ribbons, disposed in what is
called a true-love's knot!* There is no part of the
heavens where so many nebulæ and clusters

are crowded into so small a space as this " cloud."
The Nubecula Junior is a much less striking object.   It abounds more in irresolvable nebulous
light; but the nebulæ, and clusters in it are
fewer and fainter, though immediately joining to
it is one of the richest and most magnificent globular clusters in the hemisphere (47 Toucani.)"
—Fancy a firmament shaped like a lover's knot;
and, what is more, a group of firmaments manifestly related !   The assertion is assuredly war-
ranted, that when this Astronomer is restored to
England, we will obtain a wider extension of
our knowledge of the general universe, than has
befallen since the time of his great Father's discoveries.

But it is when we arrive among the almost
bewildering multitudes of unresolved systems,
that perhaps we are most forcibly struck by the
variations of their fantastic shapes.   The unresolved clusters being at depths much profounder
than the sites of the others, the sphere appro-
priated to them is of course of larger radius, and

Plate VIII.

far more capacious ; so that there is *room* for greater numbers, and also a more wonderful display of variety.   Plate VIII. exhibits a few of these curious shapes.   The annular form is sometimes met with,—one fine instance of it is in the constellation of the Lyre. The oblong sharp hoop represented in the diagram, is probably likewise a large ring, but appearing sharp in consequence of its oblique position towards us. How different utterly from ours, must be the aspects of the sky to the inhabitant of such a firmament!   The space within the ring is nearly a blank, but not perfectly so, a very thin shade of light spreading over it; so that if any intelligent eye looks from within the space upon what it may well consider its universe,—towards its sides there will be nearly an utter blank, and engirdling it round, a zone of the most dazzling lustre.   Perhaps the most peculiar of all, however, is that largest object in the Plate referred to. It has the shape of an hour-glass, or dumbbell ; the two connected hemispheres, as well as

the connecting *isthmus*, being bright and beautiful, manifesting a dense collection of stars in those
regions; while the oval is completed by two
spaces, which do not transmit a greater quantity
of light, or indicate the presence of a larger
number of stars, than the comparatively vacant
interior of one of the annuli. We are lost in
mute astonishment at these endless diversities of
character and form. But in the apparent aim
of the things near and around us, we may perhaps discern some purpose which such variety
will also serve. It seems the object, or result,
of known material arrangements to evoke every
variety of Creature, the condition of whose being can be made productive of a degree of dura-
bility; and perhaps it is one end of this wonderful
evolution of firmaments of all orders, and magnitudes, and forms, that there too, the law of
variety may prevail, and room be found for
unfolding the whole riches of the Almighty.

It was indeed a bold conception, after having

recognised the great meaning of these dim lights, and created them into magnificent firmaments— an achievement which would have filled and en- grossed most men—to undertake, besides, to compute their relative distances, and to lay down their plan. But, undaunted even by the idea of a chart of the firmamental universe, Her- schel undertook to fix what it was, within reach of his telescopes, and of course what it might be beyond them. The principles referred to near the close of last letter, apply at once to this enquiry. The power of a telescope, it will be remembered, can always be compared with the human eye,— as we know the reach of the eye, so we know the reach of the telescope ; and if that power be observed which first descries a star, a simple cal- culation will inform us, in terms of the general distances of the stars, how remote the object is. To ascertain the distance of a cluster then, note the space-penetrating power of the instrument *which first succeeds in revealing its distinct stars,* and twelve times that power will be an approxi-

mation to the required distance. Herschel, by
using comparatively small telescopes, thus fixed
the remoteness of forty-seven resolvable clusters
—ten of which were upwards of nine hundred
times more distant than Sirius. Plate IX. represents this chart, with the distances affixed.*
By larger instruments the determinations might
be greatly extended; but, though the fields are
white, the labourers have been few.—Inasmuch
as the element by which the distance is fixed,
is the *resolvability* of the cluster, we cannot
come, except by guess, at any conception of the
profundities of the milky or nebulous firmaments.
A vague guess, however, may be hazarded.
How far away, for instance, would the clusters,
whose depths we know, require to be removed,
in order to look as mere streaks, and to baffle
the powers of the best telescope? Suppose

* Of course this is a mere chart *in section* of these clusters.
They do not lie in one plane, and therefore cannot be otherwise represented than in section. It is now very imperfect
and incomplete; nevertheless it will be interesting. The
circles indicate the distances.

Plate IX

Forrester & Nichol lithog

that a cluster, ascertained as above to be of the 900th order of distances, were first seen as a whitish speck by a telescope whose space-penetrating power is 10, it is easy to calculate how far off it must be, to be first descried as a faint spot by an instrument whose power is 200. It would evidently be just 20 times farther removed from us, or at the enormous remoteness of 18,000 times the distance of Sirius. Very many unresolved clusters are undoubtedly as profound, and many still profounder in space. Calculating from the elements of a few known clusters, Herschel reaches the depth of the 35,175th order of distances, in which some of these nebulæ must lie!

And is even *this*—the UNIVERSE ? Where are we, after all, but in the centre of a sphere, whose circumference is 35,000 times as far from us as Sirius—and beyond whose circuit, infinity, boundless infinity, stretches unfathomed as ever ? We have made a step, indeed, but perhaps only towards acquaintance with a new order of *infini-*

tesimals. In our first conceptions, the distance of the earth from the sun is a quantity almost infinite;—compare it with the intervals between the fixed stars, and it becomes no quantity at all, but only an infinitesimal; and now, when the spaces between the stars are contrasted with the gulfs of dark space separating firmaments, they absolutely vanish below us. Can the whole firmamental Creation in its turn be only a corner of some mightier scheme—a larger edition, so to speak, of such a group as composes the Nubecula Major of the South—*a mere Nubecula itself?* Probably COLERIDGE is not in error :— " It is surely not impossible that to some infinitely superior Being the whole universe may be as one plain—the distance between planet and planet being only as the pores in a grain of sand, and the spaces between system and system no greater than the intervals between one grain and the grain adjacent !"

But let us not go on to bewilderment. Apart from considerations of Space and Time, we know this fact, that we are in the midst of Being, whose amount, perhaps, we cannot estimate, but which is yet all so exquisitely related, that the perfection of its parts has no dependence upon their magnitude,—of Being, within whose august bosom the little ant has its home, secure as the path of the most splendid star, and whose mightiest intervals—if Infinite Power has built up its framework—Infinite Mercy and Infinite Love glowingly fill, and give all things warmth and lustre and life—the sense of the presence of GOD!

PART II.

THE CONSTITUENT MECHANISMS, OR THE
PRINCIPLE OF THE VITALITY OF STELLAR
ARRANGEMENTS.

LETTER IV.

PROBABLE UNIVERSALITY OF PLANETARY SYS-
TEMS—RELATION OF STARS TO EACH OTHER
—DOUBLE STARS.

RETURNING homewards, through those profound abysses, to whose extremities we have
adventured, and leisurely surveying the objects
whose number and varieties struck us at first
with an absorbing and most natural astonishment, we soon start the enquiry, *What are these
clusters doing?*—What is their *internal* condition?—What their mechanisms?—And what
the nature and affections of the bodies which
compose them? It is manifest, that such investigations, in so far as we would rest them on
observation, must be confined within our own
cluster—the telescope, which has revealed the
dim lustre of others, still failing to discriminate

the peculiarities of their individual orbs; but if
we analyze the system of which we form a part,
and become familiar with the mode of *its* existence, a cautious use of the argument from analogy, will at least darkly illumine the obscurer
objects which surround it.

I. In the first place, it is of importance to
ascertain whether the stars are individually
characterised by the same leading features, or—
taking our SUN, which we know best, as a pat-
tern object—whether and how far the distinct
orbs of remote space may be accounted to
resemble him ?   The old notion that these
luminaries are of no significancy, except as
ornaments to the earth, has lost hold, I believe,
of all classes of minds, so that, assuming that
the stars are also suns, shining like our luminary, of their own perennial virtue, we may
step at once to consideration of the second or
next higher point of probable resemblance,—
are these myriads of suns encircled, like ours, by

schemes of subservient planetary worlds ? * It is singular, that the first class of stellar changes, by which the attention of astronomers was arrested, seems to refer to this relationship, and to establish at least a first presumption in favour of its reality. A number of stars have long been known to vary in lustre—increasing to a certain degree of brightness, and then waning to a certain degree of faintness—a variation found to be *periodical, i.e.* it takes place *regularly,* and within a definite time. The star ALGOL, for instance, varies regularly from the second to the fourth magnitude, and again back to the highest brightness in about two days twenty-one hours. The second star in the Lyre, goes through a periodic variation in six days nine hours. A star in the Swan varies from the sixth magnitude to absolute invisibility, and from that resumes its original brightness in eighteen years ; and similar changes, occupying different intervals, from a few days upwards to

* See note A at close of the volume.

E

many years, affect others of these remarkable
bodies.*   Now, inasmuch as the *position* of
these stars in the Heavens is wholly unaltered

* I am not certain whether another phenomenon ought not
to be ranked with these variable stars. We have now seven
or eight authentic records of the sudden appearance, and subsequent extinction of new and brilliant *fixed* stars—splendid
orbs bursting from the bosom of infinity, and after blazing for
a while, retiring slowly into their unknown remotenesses. This
phenomenon has once or twice been manifested so suddenly,
as to strike the eye even of the multitude. One of the most
remarkable instances occurred to Tycho, the illustrious Dane.
On the 11th November, 1572, as he was walking through the
fields, he was astonished to observe a new star in the constellation Cassiopeia, beaming with a radiance quite unwonted
in that part of the Heavens. Suspecting some disease or delusion about his eyes, he went up to a group of peasants to ascertain if they saw it, and found them gazing at it with as
much astonishment as himself. He went to his instruments,
and fixed its place, from which it never afterwards appeared
to deviate. For some time it increased in brightness—greatly
surpassed Sirius in lustre, and even Jupiter; it was seen by
good eyes even in the day time, a thing which happens only to
Venus under most favourable circumstances; and at night it
pierced through clouds which obscured the rest of the stars.
After reaching its greatest brightness, it again diminished,
passed through all degress of visible magnitude, and finally
disappeared. Some years after, a phenomenon equally imposing
took place in another part of the Heavens, manifesting precisely the same succession of appearances. We are quite

during their recurring changes, we cannot attribute to them an orbitual motion, or explain their varying brightness by varying distance; and, indeed, if all hypotheses purely fanciful be laid aside, there appear only two accessible explanations of this still mysterious phenomenon. Either dark bodies of considerable magnitude revolve around these stars, and by their periodical intervention intercept part of their light, or the orbs themselves rotate on axes, and at equal intervals turn towards our solar sphere parts of their surfaces which may be covered with spots, or are otherwise dimmer

baffled to account for these astonishing displays. If the bodies in question are moving in orbits, how singular, that no change of position was observable, and how tremendous the velocity which could sweep these suns in so brief an interval from a region comparatively close to us, to the invisible depths of the Heavens! Some ground of probability is furnished by a comparison of records, that the star seen by Tycho is not a stranger, but one which appeared before, passing through its mighty phasis in about three hundred years. If this be true, it ought to reappear in forty or fifty years from the present date, when, by due study, something concerning the yet strange system of phenomena it exemplifies, will probably be brought to light.

than the rest.* We know not which of these explanations is the true one; but should either be correct, it would go far to confirm the idea that the sun as a ruler of planets is not a singular orb, but a type, a representative, at least of a *class*, in the firmament. If the diminution of light is caused by the intervention of opaque revolving bodies, we have direct evidence of the reality of other planetary systems upon a scale surpassingly grand; and, if the hypothesis of *rotation* be verified, the analogy will in this case also be nearly substantiated; for—as you will learn afterwards—the rotation of an orb, and the existence of engirdling planets, are so closely connected, *that the planets may be said to spring*

* Another explanation has been proposed. The stars have been fancied of a lenticular form, so that being in a state of rotation, they would alternately present their thin and broad sides. This hypothesis is quite inadmissible, for such a shape in a body in quiescence is not known to exist. I do not say, that laws may not operate in these regions of which we know nothing, *but we are not entitled to suppose so.* No hypothesis or conjecture can be termed philosophical, which departs in the slightest degree from the rules assigned to matter by *known* laws.

out of the circumstance of rotation; the existence of planets indeed may be almost predicated
of *every* case where rotation is detected.—But
the whole problem may soon be withdrawn
from the region of surmise, and resolved in
separate instances, by the Telescope. Sir John
Herschel — of modern philosophic enquirers
facile princeps—has lately requested attention,
in the most express way, to the minute and
point-like companions of such stars as *ι* URSÆ,
α² CAPRICORNI, *α²* CANCRI, *γ* HYDRÆ, *κ* GEMI-
NORUM, &c. as in some of the cases probably
shining by *reflected* light. If these small silvery
points—lurking within the rays of their respective suns, should indeed prove to be planets, the
telescope will have performed the greatest of its
achievements ; and if, upheld by observation as
far as it can stretch, our knowledge of the physical constitution of matter shall ever enable us to
state it as a general and necessary law, that all
the orbs of space—not merely those which shine
above us, but also the myriads whose wonder-

ful clustering is seen in distant firmaments—
that each one of this mighty throng, is, through
the inseparable exigencies of its being, engirt
by a scheme of worlds—proud as ours, perhaps
far prouder;—how immeasurable the range,
how illimitable the variety of planetary existence! No wonder that our small world—a
mere nook in space—an infinitesimal item of
that mighty whole—should be incomplete and
fragmentary, silent concerning the interior of
many phenomena which are developed in it,
and containing few illustrations of much which
we desire to know in regard of the fundamental conditions of Being. The Great Book
of the Universe—that which explains the laby-
rinth and leaves no enigma, deduces its easy
expositions from the premise of the *perfect* Universe : the few stray leaves of this book which
have reached terrestrial shores, must seem
sybilline, often incoherent—speaking of laws
which enter among visible arrangements only
by their lateral actions, and whose roots are

down, far from present sight, deep in the
bosom of that all-encompassing Wisdom which
comprehends the entire system of things.

II. Rising above considerations of the indivi-
dualities of the stars, we proceed directly to the
question :—*Are the different suns isolated or
related? Does our sun, and every other orb,
retain its place, its constitution and character,
independently of other orbs ? Or, as with
minor arrangements, is there here also—even in
the great vault—some intimate connexion, some
network whose refined tracery may be pursued?*
A train of discovery has followed from these
inquiries, which might be the distinction of any
epoch, and which constitutes the proudest title
of ours to be ever illustrious in Astronomy. I
shall unfold the subject by ascending through
the different orders of known relationships.

1. About the close of last century, Sir William Herschel happened to give especial attention to a problem which had long occupied

Astronomers—the problem which proposes to determine the distance of the stars, by noting the amount of their apparent displacement, when viewed from opposite ends of the Earth's orbit. The difficulty, next to insurmountable, of resolving this problem by observations on single stars, induced him to revive an idea originally thrown out by GALILEO, that the end might be achieved by watching the distances and position of two stars almost in apparent contact.* A few such objects had long been known to exist—objects appearing single to the naked eye, but separating into two stars when

---

* These objects are exceedingly beautiful, and fortunately some of them may be decomposed by telescopes of ordinary power. There is one in the constellation of the Great Bear, which a common glass will separate for you; it is the star ζ.

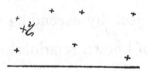

Even to the naked eye ξ appears double, but that is not the phenomenon. It is the *larger* of the two, which may be decomposed, and to which I refer you.

examined through adequate telescopes; and the commonly received explanation of them was, that they were two stars, not in contact, not even in proximity, but simply in almost the *same visual line*, the one probably very *far behind the other*. I do not know if Herschel began by doubting the truth of this explanation, although its accuracy was vital to the success of his first proposal; but without long delay circumstances led him to conclude it wholly untenable, and he proved so by a process of ratiocinative observation of a very pregnant kind, which is still one of our most brilliant examples of philosophic enquiry into the remote unknown.

With certain exceptions of no consequence at present, the stars—as I have stated—may be held to be nearly equally distributed in the firmament; and upon this hypothesis of equal distribution, it is easy to calculate approximately how many, or at least what *proportionate number of the whole*, would be placed—the one so nearly behind the other; and we find that although the hypothesis of equal distribution were modified much more than appearances require, the numbers thus determined on would not be materially increased. But the actual number of DOUBLE STARS, which revealed themselves to Herschel, *immensely surpassed what any calculation like the foregoing could authorize*; upon the ground of which anomaly, he boldly averred, that, connected with these multitudes of joint orbs, there are indications of a distinct and *systematic arrangement*. I think I can easily place you where Herschel then stood. Suppose a number of peas were thrown at random on a chess-board, what would you expect? Certainly, that they should be found occupying irregular or random positions; and if, contrary to this, they were, in far more than average numbers, arranged by *twos* on each square, it would be a most natural inference that here there was *no random scattering;* for the excessive prevalence of the binary arrangement would indicate forethought, design,

system. " Casual situations," says Herschel,
" will not account for these multiplied phenomena; and consequently their existence must
be owing to *some general law of nature.* Now,"
he continues, " as the mutual gravitation of
bodies toward each other is quite sufficient to
account for the union of two stars, we are
authorized to ascribe such combinations to that
principle." You perceive he has at once started
to the fact, that a double star is not always *one
star seen behind the other,* but for the most part
two stars in actual proximity—in *systematic and
regulated union;* and in the fullest confidence in
that ORDER which pervades all Nature, where
every arrangement has a cause and meaning,
he has also asserted that this union *subsists
through energy of some general law.* It needed
little more to enable him to foreshadow a memorable truth.—Gravitation, as Herschel well
knew, if it accounts for the proximity of bodies,
does not account alone and of itself, for the
stability of systems ; and, believing that every

system in nature is so far stable, he proceeded
to endow these stars—prior to observation of
the fact—with ORBITUAL MOTIONS, asserting
that the one must revolve around the other, as
the earth around the sun, so that the attractive
energy might be retained in check.*   In a
paper which I account one of the few very remarkable memoirs among the records of observant philosophy, this great man fully unfolds
the *necessity* of these yet undiscovered motions,
and ventures to chart their plan.   How grateful must have been the fulfilment of the predic-
tion, and how fortunate that by the veteran's
own labours, and those of other observers,
among whom his illustrious son stood foremost,
these expectations were amply realized, ere—
covered with honours—he descended into the
tomb !   Dwelling on this instructive incident, I

* Note A, at close of the volume, which I hope you have
perused, will explain how the orbitual motion of the planets
checks the sun's attractive influence. Quite the same principle
is applicable to the Double Stars.

have seemed to perceive more clearly than formerly, the difference between genius and mere
industry, between the plodding accumulator of
facts and the man whose eagle eye, wherever it
wanders, can pierce through the surface and rest
on the fine internal relations of Nature. Philosophy of this sort necessarily involves the spirit
and power of prophecy; and, so it has been that
few great truths have ever arisen upon our
world, of which the meditative and illustrious
did not, long ages before, trembling between
Hope and Fear, venture to hail the Aurora.

Ideas of the promise of the foregoing could
not be cast fruitlessly upon the world. The
endeavour to verify them has indeed constituted
the main work of astronomy from that time till
now; and the nature of the expectation they
excited, has determined the course and character of observation. Before any thing definite
could be established concerning the systematic
nature of these double stars, the apparent distances and relative position of the two con-

stituents of each duplex system required to be
fixed, as they stood at a given date, and compared with determinations of the same elements
at a subsequent date, separated from the previous one by the interval of a considerable
number of years. If the two measurements
were the same, it might be concluded that no
motion had supervened; and if, on the contrary,
motion had ensued, not only would the fact be
indicated by a change of the observed elements,
but the *nature* of that alteration, would enable
an experienced mind to determine the *kind* of
motion, or the kind of curve in which the bodies
had moved.* The first step of the enquiry is thus,
to collect all these first elements possible, to sur
vey the whole heavens in search of double stars

---

* Are you doubtful of our powers of measuring quantities so
small—so evanescent as the relations of these close bodies?
Art is still very far from having expended its full resources on
micrometers, although it has done much. It is a common
thing to measure a quantity no greater than the 1-9000th part
of an inch; but I have been told by persons conversant with
the dividing engine, that the 1-20,000th part may be made
appreciable!

and to tabulate their existing condition. This, as it were, is stretching out our base line—clearing the ground; an employment always laborious, but not beyond the enthusiasm of astronomers. A fresh life seemed suddenly to come over science —something like the excitement diffused through Europe when Columbus discovered, in the far west, new and immeasurable worlds resting amid the formerly void and mysterious ocean. Herschel himself completed a list of 500 excellent measurements; and I record with delight the names of those who have most worthily followed him. In 1824, SIR JAMES SOUTH and SIR JOHN HERSCHEL, produced a catalogue of 380 stars, whose distances and angles of position, they had jointly fixed with admirable precision; and their paper, which is a noble memorial of friendship, will always be a model in this kind of enquiry. South followed it up by a distinct catalogue of 480; and Herschel, now also observing apart, has completed a list of upwards of 3300 of such bodies, accompanied by the requisite determinations. To the names of these

ornaments of British astronomy, it is most grateful to add that of STRUVE of Dorpat, who, aided
by that *nonpareil* of mechanism, the Equa-
torial of Frauenhofer, has succeeded in arranging a masterly catalogue of no less than 3000
double stars. Undoubtedly many objects are
common to the lists of Herschel and Struve,
but these numbers will give you an idea of the
advances we have made.   With all this, however, even in the northern hemisphere, we are
only entering the field, for in the profound vault
there are probably myriads. To discover Struve's
3000, not more than 120,000 stars were examined, and what are these to entire contents
of the skies? Sir William Herschel computed,
that in the milky way 116,000, or more than
nine-tenths of this whole number, have passed
under his review during a quarter of an hour's
observation; so that in undiscovered systems,
the Heavens are still marvellously rich, and
mechanisms, more singular perhaps than any
yet known, may lurk within these masses, now
only revealed by their obscure and aggre-

gate light. The ground is thus cleared only over a trifling territory,—the measurement of the base is but begun; and observers are solicited to the labour, and implored to shrink from neither watchings nor toil! The enduring glory of TYCHO, whose name is as a star, whispers a sweet promise of immortality to him who will here be a victor over the unknown; if indeed external reward is needed, or aught else than the pleasure of unwinding the harmonious arrangements of the Universe, and the first vision of their magnificence. In search of magnificence, it is true, we need not wander far,— witness the fields which encircle your home, the blade of the modest grass which adorns them; but the Heavens are *fresh*,—familiarity has not left its footprint on their untrodden floor. In the silence of warm midnight, that noble curtain stretched out above me, and the idea present and impressive, of its orbs obediently pursuing their stupendous paths, I confess there is a solemnity which sometimes falls upon the

F

spirit, not unlike the feeling of the Patriarch, when he heard that low rushing wind, believing it to be the audible footsteps of his Creator!

I have said that these catalogues, are so to speak—only measurements of a base—first or preparatory surveys; but, while in general they are so, they contain consecutive determinations with regard to a few stars, which Sir William Herschel early observed; and these have already enabled us to draw the most decisive conclusions— proving that the general ideas of our philosopher are not illusory, but sound interpretations of the plan of Nature. *A considerable number of systems have distinctly yielded the phenomena of their revolutions,*—a fact which I shall place before you in its highest evidence, as I desire that you henceforth regard the orbitual motions of suns, to be no less indubitable than the fact of the revolution of the earth. Let me detail, therefore, the full grounds of our belief in two instances, which are patterns of all others; I choose ξ Ursæ Majoris and Castor. The following Table gives a synoptical view of the series of observations made upon the former star, along with the names of the observers, and dates of observation.

ξ Ursæ.

Epoch.	Position.	Distance.	Observers.
1781.97	503°.47′	—	Sir W. Herschel.
1802.09	457°.31′	—	Do.
1809.08	452°.38′	—	Do.
1819.10	284°.33′	2″.56	Struve.
1820.13	276°.21′	—	Do.
1821.13	268°.48′	—	Do.
1822.08	262°.39′	—	Do.
1823.29	258°.27′	2″.81	South & Herschel.
1825.22	244°.32′	—	South.
1825.25	—	2″.44	Do.
1826.20	230°.42′	1″.73	Struve.
1827.27	228°.17′	1″.79	Do.
1828.37	224°. 1′	2″.01	Sir J. Herschel.
1829.02	219°. 0′	2″.	Do.
1830.18	211°.25′	—	Do.
1830.24	—	2″.08	Do.
1830.98	200°.54′	2″.39	Do.

Be not alarmed at this array of figures—its significancy is easily understood. A diagram will at once explain what is meant by its important column—the list of *angles of position*. Suppose one of the stars in the centre of the subjoined circle, or at A; and let A o° be the line

in which the two are conceived to make *no* angle
with each other—the primary or normal direction to which all others are referred.   As the
second star moves out of that line, whether to
the east or west, the line joining the centres
of the two, will manifestly diverge from the first
line A *0° or form an angle with it*; and the
quantity of the circle which the new line cuts
off *measures this angle*.   The angles of position in the second column then simply indicate *how much of the circle*—counting from *0°*—
*is cut off in each case, by the line joining the
centres of the two stars ;* so that we know exactly
towards what part of the circle the line in question must have pointed at the different epochs
in the table; and, if we mark these places

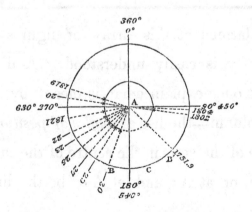

on the circumference, we will obtain a graphic
representation of the changes which have supervened on the relative position of the two
stars during these fifty years. The representation is nowise enigmatical. The first recorded
angle being 503°, the second star must have
been originally found by Sir William Herschel,
in a direction from the first, of a line drawn from
A, towards the date 1781.9, as marked on the diagram. In 1802 and 1804, again, we discover it
in new directions, pointing towards those numbers on the circumference of the circle,—a change
which cannot possibly be explained save on the
hypothesis of *the relative motion of the constituents of this star ;* and in subsequent years—as
the figure continues to represent, it migrates
through all the positions noted, moving almost
around the circle, or, in other words, with
exception of the small reverse arc, B C B', *performing a complete orbitual revolution.* The surprising fact is thus demonstrated—placed beyond
all reach of doubt—that such motions exist, or

that SUNS REVOLVE AROUND SUNS, and an idea is introduced which alters the entire aspect of our Astronomies. It is computed that the arc B C B' will likewise have been passed over before the year 1840, the period of the system being somewhat upwards of 58 years. The other binary star I would refer to, is CASTOR.

CASTOR.

Epoch.	Position.	Distance.	Observers.
1719.84	355°.53'	—··	Pound and Bradley.
1759.80	326°.30'	—	Bradley and Maskelyne.
1779.84	302°.47'	—	Sir W. Herschel.
1791.64	295°. 6'	—	Do.
1795.96	283°.54'	—	Do.
1802.04	281°.22'	—	Do.
1813.83	272°.52'	—	Struve.
1816.97	270°.—	—	Sir J. Herschel.
1819.10	269°.36'	5".48	Struve.
1821.21	267°. 7'	—	South and Herschel.
1822.10	—	5".36	Do.
1823.11	264°.59'	—	Do.
1825.26	263°.18'	4".77	South.
1826.29	262°.38'	4".42	Struve.
1828.60	261°.52'	4".64	Sir J. Herschel.
1829.88	260°.58'	4".52	Do.
1830.52	259°. 1'	—	Do.
1830.60	—	4".68	Do.

Castor has manifestly a much slower motion than ξ Ursæ. Notwithstanding the greater extent of period, no less than 117 years, which contain the changes above recorded, the alterations are far less signal. The sphere of motion—represented, as in the former instance, is comprised within the arc of the subjoined circle, indicated by the commencing and terminating dates; so that although the reality of a

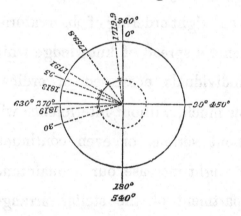

motion is fixed no less decisively, only a small portion of the course of this star has yet been witnessed; and we know from other considerations that its full period occupies more than 250 years.—The idea that labours, as long and continuous as those which have brought to light

the relationships of the two systems now in-
stanced, must be expended on every individual
in Struve's or Herschel's catalogues, is almost
appalling ; yet, unshaken in my confidence
that by far the greater number of these multitudes, are revolving *systems*, I do not hesitate
to assert that, if we would explore this part of
the Universe, *it must be done.* The solution of
the oppressive difficulty is to be sought in the
multiplication and right ordering of observatories,
and in that general spread of knowledge which
will inform individuals possessed of ordinary
telescopes, how much, without interference with
business, without serious or even continuous
exertion, they might increase our acquaintance
with this department of the stellar arrangements.

The following table, divested of details, will
make you acquainted with the periods of a few
other known systems : ζ Cancri and η Coronæ,
have both performed *complete revolutions* since
observation began on them.

Stars' names.		Periods.
η CORONÆ,	.	43 years.
ζ CANCRI,	.	55 ...
70 OPHIUCHI,	.	88 ...
σ CORONÆ,	.	287 ...
61 CYGNI,	.	452 ...
γ VIRGINIS,	.	629 ...
γ LEONIS,	.	1200 ...

The fact of motion is established in a considerable number of other cases, but hitherto we
have not approximated to their periods. It is
understood, however, that Sir John Herschel
has detected among the multitudes in the
southern skies, several very remarkable binary
stars of periods still briefer than any on our
records ; we have indeed reason to expect this
philosopher's return with impatience ! While
referring to these remote Heavens, let me not
overlook Mr DUNLOP of Paramatta, whose
catalogue first introduced us to the double stars
and clusters which swarm in them ; nor must

Sir THOMAS MACDOUGALL BRISBANE be here forgotten—himself an admirable and zealous observer—through whose munificence and enterprise, that observatory was planted in Australia, and British science made co-extensive with British dominion. Gracefully does the laurel due to such actions adorn the green autumn of life !

Hitherto I have spoken merely of the fact of orbitual motion. The nature of the curves or paths in which the bodies move, demands further investigations, which have already yielded a most important general truth. At first sight, it seems easy to detect these paths, the problem simply requiring that we set off—as represented in the diagram of page 84—distances from the central body O, corresponding to each angular position, and taken from the third column of the tables. Unfortunately, however, we cannot depend upon our observations of distance; the slightest error vitiates

the whole, and the best observers commit at
present more than slight errors in such estima-
tions.   For instance, the star 70 Ophiuchi
would, if so judged, have an orbit like the following, through which no imaginable contin-

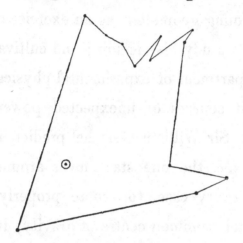

uous curve could pass.   There is a supreme
necessity for the rejection, *in toto*, of all recorded distances, and that reliance be placed
on the angles of position alone.   Acting virtually on this plan, SAVARY evolved, by a long
but ingenious analytical process, the orbit of
ξ Ursæ, and Professor ENCKE of Berlin applied similar methods to the case of 70 Ophiuchi ; but the honour of summarily overcoming

the whole difficulty—of giving a general and
pliable method, distinguished no less by its
facility than its extreme beauty, by which the
required curve may be charted, even from partially erroneous elements, is due to Sir John
Herschel.   Young geometers, as an exercise of
taste, ought to study his memoir; and cultiva-
tors of any department of experimental physics,
will find in it sources of unexpected power.
The curves, as Sir William Herschel predicted,
are *elliptical; i. e.* the one star moves around
the other, in every case, (or, more properly,
each around their common centre of gravity) in
an oval or elliptic curve—*precisely the curve
which is described by the Earth and other planets
in their revolutions around the Sun.* Uniformity
of this sort is exceedingly remarkable—*it points
to some common cause;* in other words, to the LAW
OF GRAVITATION, which the nature of this curve
enabled NEWTON to detect as the first principle
of planetary order.   Gravity has often been sur-
mised to be universal; at all events, we have

now stretched it beyond the limits of the most eccentric comet into the distant intervals of space. Every extension of its known efficacy manifestly increases, in accelerating ratio, the probability that it is a fundamental law or principle of matter; but although it should somewhere fail, it is still a type of the mode of the constitution of things;—it will lose its universality only through the preponderating efficacy of still profounder powers. Judged in this true light, the vastness of Creation is comprised within a mighty plan; and we, standing on this little world, can gaze around on its majesty, and note its stupendous changes in peace—knowing that there is no hazard or caprice in Mutability, but only the stern and steadfast power of Law, through which events roll onward to their destiny.*

Numerous supplementary remarks are suggested by the phenomena of these curious

* See note B, at the end of the volume.

systems ; but they are scarcely within my province. To one additional circumstance alone will I allude, in conclusion—partly because of its inherent interest, and partly because it unquestionably points to some important, although still undefined fact, relative to the physical constitution of the celestial bodies. The *light* of the stars is by no means uniform,—the ray of Sirius for instance differs, not merely in intensity but in *kind*, from the ray of Vega; and in countries where the atmosphere is less humid and hazy than ours, the difference is striking to the naked eye—one star shining as an emerald, another as a ruby, and the whole Heavens sparkling as with various gems. This attribute of variety of colour also characterizes the double stars; but the remarkable thing is that in many instances, where one star is of one marked colour, its companion is of the opposite. Instances abound in which a red and a green star are associated, or a yellow and blue. When the stars are of different degrees of brilliancy, this

contrast may originate in an optical delusion—
in that tendency which disposes the eye, when
gazing on any bright colour, to endow fainter
objects near it with the opposite or complemen-
tary colour by way of relief; but the explanation
is not universally borne out, inasmuch as many
couples in precisely similar circumstances, show
no such contrast. Sir John Herschel was at
first decidedly inclined to attribute the phenomenon to an actual difference of colour, and although he has since—perhaps on good grounds—
half relinquished that conclusion, we have the
acquiescing testimony of Struve, founded on
observations with the Dorpat Telescope, whose
clearness has never yet been rivalled, so that we
may not absolutely part with the early pleasing
speculations of the British astronomer. " It
may easier be suggested in words," says Sir
John, " than conceived in imagination, what
variety of illumination, two stars—a red and a
green, or a yellow and blue one, must afford a
planet circulating around either ; and what cheer-

ing contrasts and grateful vicissitudes, a red and
a green day for instance, alternating, with a white
one and with darkness, might arise from the presence or absence of one or other, or both, from
the horizon!"   The mention of planets, starts
speculations, equally curious, and much less hypothetical. If, in consequence of a law at whose
probability I have already hinted, small encircling
worlds are a necessary appendage of each sun,
what a field of various and strange life is opened
by the idea of Spheres of the nature of ours
with two suns—having sometimes one, sometimes
both, and sometimes neither, burning in their sky!
All the products of the material constitution of this
earth, the character of its living families, perhaps the action of its magnetic and other influences, are co-ordinated and adjusted to the regular succession of night and day, or to the supply
and absence of solar heat. No such families
then, none bearing other than remote analogies
to ours, can exist in planets, engirdling those
double suns. They, too, are surely the abodes

of beauty and harmony, but their features are
hidden from man,—perhaps for ever. And who,
after all, would grieve, although there be some
enclosed spots—quietudes, in Creation, which will
be unexplored, unpenetrated for ever; who that
has felt the soft healing of Evening can regret
that even in the intellectual world, there are regions into which faintness and weariness may
sometimes flee, and take shelter and repose, away
from the scorch and glare of oppressive light!
Sweet and inviting mysteries—among whose
gentle shadows, Hope and Fear, and all unnamed yearnings tremblingly advance, and find
or fashion for themselves, images of purity, con-
victions of immortality, vistas of a long life to
come, through which the soul may wander freer
and greater than now, having gained the privilege by virtue!

LETTER V.

TRIPLE STARS—MORE COMPLEX RELATIONSHIPS
—EXISTENCE OF LARGE GROUPS—CONJEC-
TURES RESPECTING FIRMAMENTAL SYSTEMS.

2. GUIDED by the Genius, whose prophetic eye pierced the obscurity, which, until then, had concealed the arrangements we have just unfolded—we advance cautiously, but without dread, to take cognizance of still higher schemes.— The arguments which induced Herschel to pronounce on the connexion and motion of the constituents of binary systems, penetrate much farther, and intimate as a general law, *that every cluster or unusual aggregation of orbs must be systematic, and probably united by common motion.* If it is unlikely that the principles of random scattering would produce numbers of DOUBLE

STARS, it is plainly as unlikely that *any* TRIPLE
or QUADRUPLE BODIES should be found in the
whole sky ; and this holds even where the stars
are more separated, as in the case of the six principal constituents of the Pleiades, against whose
fortuitous aggregation within that space—as
Michell long ago calculated—there is a balance
of probabilities of 500,000 to 1.   The presence
of a great law thus lays it on astronomers as a
command that they watch these higher systems,
take their measurements with every minute-
ness, and transmit them to posterity.   The
results must be brilliant, and they are already
foreshadowed.   In a triple star in CANCER
(ζ), we are certain of a common motion, in
which three suns seem to revolve around a
common central point ; and in another—ψ CAS-
SIOPEIÆ — one sun probably revolves around
a second, while the two in union—a sun and
an associated sun, circulate around the third.
The quadruple star ε Lyræ, is in all likelihood, a quadruple system, whose motions are

exceedingly complex and singular—perhaps as
follows,—the star A revolving around B, the

star C around D; the system of A and B revolving around a point O between B and C,
and the system of C and D being carried around
the same point in another orbit. If we revert
to our supposed law of necessary planetary
existence—if the orbits of dependent worlds
are intertwined around these four luminaries,
we must indeed have strange systems in space,
and mechanisms of a complexity which compels
our boasted powers of analytical calculation, to
confess themselves in the very infancy of progress ! I doubt not there are schemes still more
intricate than the foregoing—for instance, *double
triple* stars, as with the chief group of σ Orionis;
nor, although positive discovery is not achieved,
are indications wanting of activity among the
larger clusters. The lost Pleiad—the sorrowing

Merope—is matter of song; perhaps the Mythology speaks of the retirement of a star formerly visible: and "of late years a fact much
less apocryphal, has pointed to a similar and
striking phenomenon.   In the beautiful tra-
pezium in Orion, only *four* stars appeared to
our best instruments from the time when the
telescope took on its power; but Struve had
scarcely seen it with his great refractor, until he
discovered a *fifth*.   He lost no time in notifying
the discovery to Sir John Herschel, who confirmed it with his twenty feet reflector, stating
besides, that if the star had been previously
visible, he could not have missed it, on occasion
of his sketching with great care that outline of
the Nebula, to which I shall soon require to
refer.   These indeed are isolated facts, but they
are to be interpreted in connexion with an important and far reaching principle; and astronomers will not be justified in yielding them only
an ordinary attention.—When one thinks of the
vast field opened by such conjectures—of the

little hitherto accomplished, compared with what remains to be done—of the deep mystery still hanging over almost all of the vast skies— there is apt to supervene a despondency, a hope- lessness that the handwriting which is on them, will ever be interpreted. But we take encourage- ment from the aspects of the times. Astronomy is not now in that stage of its history in which only a few men in a century would consent to wear out a long and healthful age in examining the Heavens. Observers of the first capacity and becoming ardour, are yearly multiplying; while adequate instruments, through the advance of Art, grow more accessible. Doubtless, with all advantages, we, of this time, may do little more than roughly chart the boundary lines, and it may be, fix down the prominent points of the landscape ;—the filling up and mapping of the details constitute the harvest of the future. But how soon may that future come! The wheels of time are revolving rapidly—truth mingling with truth, as light gathered into a

focus. Alike within and around us, events suc-
ceed without the usual interval, nor is astronomy
unaffected by the general acceleration. The
knowledge of what the Heavens are boding may
not be long deferred; if we, in present times, in-
dustriously act our part, much still unintelligible
will become plain to the generation whose
buds at this moment are the spring tidings of
the world—the generation now pressing on us,
and to which we must yield the stage.

3. If the investigations now shadowed out,
shall, in conformity with Herschel's conjectures,
establish the general principle that all minor or
peculiar clusters have a systematic character, a
great light will undoubtedly be thrown upon
the forms of the existing Universe; yet it
cannot be overlooked, that the existence of
close groups is not the leading feature of our
firmament. Those masses of dispersed suns,
which chiefly compose it, have probably also
mutual relations; and happily in one case we

have proceeded a brief way in the task of showing their reality. The comparison of catalogues
of the stars, taken at different epochs, leave it
unquestionable that a considerable number of
bodies not connected in any close group, appear
to have *proper motions*,—that is, their places
change annually by extremely minute quantities : out of a catalogue of 314, lately presented by Mr Baily, I select POLARIS, SIRIUS,
PROCYON, POLLUX, ARCTURUS, CAPELLA, and
ATAIR, as illustrations. Now, these stars—
and it is the same with the whole three hundred and fourteen—seem to have no particular
locality. They are scattered on all sides of us
with apparent indifference,—a circumstance
which at one time led Sir William Herschel to
suppose that their changes might result, not
from proper motion, but—as with the revolution
of the Heavens—from *a motion on our part*,
from the fact of our sun along with his planets
advancing towards the constellation HERCULES.
If, as is exceedingly likely, our sun has an

extensive motion of translation, and with his dependent train is sweeping toward some remote point of space, or circling around some balanced centre of attraction, there is indeed no doubt that the effect of his motion will be mixed up in what will appear the proper motions of the other stars; but it has been distinctly shown by BESSEL—perhaps the first practical astronomer of modern times—that the actual changes of the stars cannot be accounted for on this sole principle, and that we must attribute to a great number of luminaries a decided proper motion. We are surrounded, then, by a considerable number of orbs widely separated from each other, which yet have mutual affections of a nature we have not comprehended, but whose reality is indisputable; and *our sun, being in the midst of this group, may be held as belonging to it, and mingling with its relations.* We know not how far the influence of this relationship may extend;—perhaps the planets attending the sun may owe to it something of the physical variations to which they are subject. The recent conjecture of a continental analyst is not to be summarily rejected or overlooked in a philo-sophical induction, — that a degree of those changes of temperature which the earth has undergone since life appeared in it, and because of which our northern climes were one day capable of harbouring the palms and gigantic ferns of the tropics—may have supervened in consequence of our gradual translation into chiller regions of space.

These motions are not comprehended within Herschel's principle; and as I have said, they seem to point to the constitution and inter-nal affections of a large group, but still of a mere *group*, and do not reveal otherwise than *secondarily*, the nature of the mechanism of the complete or entire firmament. Since the observed motions affect only a particular number of bodies, or a *district* of stars—I see no reason to attribute to them a wider signifi-cancy; and the very capricious figure of the

cluster to which we belong sustains our idea.
Turn again to Fig. 1, Plate III. :—It is extremely irregular, and yet may present only
the leading peculiarities of a cluster like ours;
while a minuter representation might indicate
lines or partitions—*groups* in fact—which are
invisible through effect of distance.   Not only
might the ring have less appearance of uniformity, but the central mass itself, within something like which our sun is located, might be
divisible into distinct spaces—self-contained subordinate clusters, whose aggregation makes up
the whole.   This hypothesis is supported by an
inspection of our ring.   The milky zone of our
firmament is not uniform.   It is rather a succession of bright spots, generally separated from
each other by a comparatively dark line or space,
manifesting a succession of clusters, which, no
doubt, are mutually related and bound together
in one whole, but which likewise have a peculiar
internal systematic character.   The naked eye
easily detects this feature of the milky way.

Plates X. and XI. are eye sketches of part of
this zone by Mr Dunlop, showing its aspects in
the southern hemisphere, which you will observe
precisely correspond with my description of its
northern regions.   Unhappily we yet know
nothing of the mechanism of these great clusters, and can only conjecture that the motions
of that group within which our sun is imbedded,
may be somewhat analogous.   " How much,"
says Sir John Herschel, " is escaping us ! How
unworthy is it of them who call themselves
philosophers, to let these great phenomena of
nature—these slow but majestic manifestations
of the power and glory of God—glide by un-
noticed, and drop out of memory beyond reach
of recovery, because we will not take the pains
to note them in their unobtrusive and furtive
passage, because we see them in their every-
day dress, and mark no sudden change, and
conclude that all is dead, because we will not
look for signs of life ; and that all is uninterest-
ing because we are not impressed and dazzled."

Plate X.

Forrester & Nichol lithog

Plate XI.

Forrester & Nichol lith.

" To say indeed," he adds, " that every individual star in the milky way, to the amount of
eight or ten millions, is to have its place determined, and its motion watched, would be extrava-
gant; but at least let samples be taken—at least
let monographs of parts be made with powerful telescopes and refined instruments—that we
may know what is going on in that abyss of
stars, where at present imagination wanders
without a guide !"

4. We approach the land of clouds and doubt
—the solid ground of fact and observation is
rapidly retiring from below us. Of the nature
of the relations of these subordinate groups, or
of the binding or compacting principle—so to
speak—of our singular Firmament, considering
it as a *whole*, I can inform you nothing. The
more capricious class of external clusters equally
mocks our imaginings; although the application of Herschel's principle fully assures us, that
these vast masses, are not grouped together by

chance or at random, but that every such union
of stars indicates law and system, as clearly as a
binary or ternary aggregation. The only light we
find among these spaces is a welcome gleam of
evidence that Nature there also is uniform, since
the simpler Firmaments manifest by their shapes
the prevalence of an *internal attractive power*
—like that whose effective presence bestows
consistency on our own subordinate clusters.
Notwithstanding the apparent caprice of many of
the firmaments, the *round* or *globular* structure is
the general or favourite one, as you may judge
from the representations in Plates I., V., VI.;
and in most of these round clusters, there is also
a *strongly marked increase of light towards the
centre,* much more than would arise from the
circumstance of our there looking through the
deepest part of the group, and thereby seeing
at once the greatest number of its stars. This
latter phenomenon decidedly indicates *compression* in greater or less degree, nor is it confined to masses having the perfectly spherical

figure. " There are besides," says Sir William
Herschel, " additional circumstances, in the ap
pearance of *extended* clusters and nebulæ, which
very much favour the idea of a power lodged
in the brightest part. Although the form of
these be not globular, it is plainly to be seen,
that there is a tendency towards sphericity, by
the *swell* of the dimensions the nearer we draw
towards the most luminous place, denoting as it
were a *course* or *tide* of *stars*, setting towards a
centre. And, if allegorical expressions may be
allowed, it should seem as if the stars, thus
flocking towards the seat of power, were
stemmed by the crowd of those already as-
sembled, and that while some of them are successful in forcing their predecessors sideways,
out of their places, others are themselves
obliged to take up lateral situations, while all
of them seem eagerly to strive for a place in
the central swelling and generating spherical
figure."

It were presumptuous to adventure farther.
The mere starting of questions, on the ground
of observation, as to the laws of the present
distribution of the different firmaments, must be
postponed to a future day;—let us rejoice that
it has required so brief an interval since utter
darkness, to enable the mind to raise itself to
the height of deeming such subjects cognoscible.
If coming times be as fertile as the past, and
produce and educate new Galileos and Herschels, they must resolve every difficulty, and
remove the veil from all that is obscure. What
triumphs—what delights—are awaiting us! Yet
once more shall new views dawn upon mankind
—yet once more will some favoured eye first
track a vast unknown! Anticipating such
achievements, one almost participates in the
elevation of the man who will realize them; and
that mood leads us best to estimate the worth of
those who have gone before, and conquered what
we now enjoy. It is indeed an attractive thought
—to revert to the period when the forty feet

telescope first interrogated the profound heavens!
The enthusiastic observer in the act of discovery,
rises before the imagination, and the peace of midnight and the beauteous twinkling of the stars;
as also that other feature which characterised and
farther elevated the scene. The Astronomer,
during these engrossing nights, was constantly assisted in his labours by a devoted maiden
Sister, who braved with him the inclemency
of the weather—who heroically shared his pri-
vations that she might participate in his delights—whose pen, we are told, committed to
paper his notes of observations as they issued
from his lips; " she it was," says the best of
authorities, " who having passed the nights near
the telescope, took the rough manuscripts to
her cottage at the dawn of day, and produced a
fair copy of the night's work on the ensuing
morning; she it was, who planned the labour of
each succeeding night, who reduced every observation, made every calculation, and kept every
thing in systematic order;" she it was—MISS

H

CAROLINE HERSCHEL—who helped our astronomer to gather an imperishable name. This venerable lady has in one respect been more fortunate
than her brother, she has lived to reap the full
harvest of their joint glory. Some years ago
the gold medal of our Astronomical Society was
transmitted her to her native Hanover, whither
she removed after Sir William's death; and the
first of our Learned Societies has recently inscribed her name upon its roll: but she has been
rewarded by yet more, by what she will value
beyond all earthly pleasures; she has lived to
see her favourite nephew, him who grew up
under her eye unto an astronomer, gather around
him the highest hopes of scientific Europe, and
prove himself more than equal to tread in the
footsteps of his father.

PART III.

THE ORIGIN AND PROBABLE DESTINY OF THE PRESENT FORM OF THE MATERIAL CREATION.

LETTER VI.

THE NEBULÆ.

THE disclosures of the telescope are now before us,—the entire perspective of Modern Astromomy. Can we comprehend its wonders? Are its arrangements a fixed thing—a mere passing show—or are they results of a pre-existing state, and germinant of something future? These questions warn me, that again we break new ground, and enter on speculations, perhaps the most adventurous which have yet engaged the reason of Man.

Astronomy has recently been obliged to recognise a Matter—or rather a modification of Matter, wholly distinct from stars—a thin and filmy substance diffused through the stellar intervals, and spreading over regions so immense, that its magnitude or the space it fills, is absolutely inconceivable. It unquestionably becomes us not to admit an element so remarkable, and which, if real, must perform important functions, and materially affect our general views of things—until its claims have undergone the severest scrutiny ; and as I am desirous to convey to you full power of judging for yourself, I will here minutely follow the process of thought, by which Sir. William Herschel— only, however, at a comparatively late period in the course of his researches—was, slowly and almost reluctantly, led to the conviction of its reality.

In his earlier inquiries, Herschel was inclined to consider all the faintly illuminated spots in the heavens, as clusters so remote, that only their general illumination, and no individual object could be seen; and the inference, so far from being constrained, seemed to result from his whole previous experience. On viewing the

heavens, for instance, with a seven-feet reflector,
while many distinct clusters were revealed to
him wholly invisible by the naked eye, a great
number of new illuminated spots were also
visible.   Now, on applying the ten feet telescope, a proportion of these last were at once
resolved into stars—others formerly of a milky
hue put on the resolvable aspect ; that is, they
seemed like a distant handful of glittering dust ;
and although many retained their former irresolvable appearance, what more natural than to
refer their continued intractability to their still
greater distance?   " When I pursued these
researches," says our Astronomer, " I was in the
situation of a natural philosopher, who follows
the various species of animals and insects, from
the height of their perfection down to the lowest
ebb of life ; when arriving at the vegetable king-
dom, he can scarcely point out the precise boundary where the animal ceases and the plant begins,
and may even go so far as to suspect them not
to be essentially different.   But recollecting him-

self, he compares for instance one of the human
species with a tree, and all doubt upon the subject vanishes before him.   In the same manner,
we pass by gentle steps from a coarse cluster
down through others more remote, and therefore
of a finer texture, without any hesitation, till we
find ourselves brought to an object such as the
NEBULA in ORION, when we are still inclined to
remain in our once adopted idea of stars exceedingly remote, and inconceivably crowded, as being
the occasion of that remarkable occurrence.   It
seems, therefore, to require a more dissimilar
object to bring us right again.   A glance like
that of the Naturalist who casts his eye from
the perfect vegetable to the perfect animal, is
wanting to remove the veil from the mind of
Astronomers."

The object which broke in upon Herschel's
previous continuity of inference, was a *nebulous
star*—a perfect star with a halo or dim atmosphere around it,—such an object as is represented in Fig. 1, Plate XIX.   I transcribe the

record of the observation, and his subsequent remarks. After noting the elements which fix the star's place, he says, " A most singular phenomenon ! A star of about the eighth magnitude with a faint luminous atmosphere of a circular form, and about 3′ in diameter. The star is perfectly in the centre, and the atmosphere so diluted, faint and equal throughout, that there can be no surmise of its consisting of stars." Herschel arrived at the latter positive conclusion as follows. " In the first place," says he, " if the nebulosity consists of stars appearing nebulous because of their distance, which causes them to run into each other, what must be the size of the central body which, at so enormous a distance, yet so far outshines all the rest ? In the next place, if the central star be no bigger than common, how very small and compressed must be the other luminous points which send us only so faint a light ? In the former case, the central body would far exceed

what we call a star; and in the latter, the shining matter about the centre would be too small
to come under that designation.   Either, then,
we have a central body which is not a star, or
a star involved in a shining fluid of a nature
wholly unknown to us."   The latter alternative
may, at first sight, appear the strangest and
the most remote, yet it is the one to which the
balance of probability manifestly inclines.   And
our judgment rests upon this,—the nebulous
fluid, supposing it to exist, could not become
known *under any other aspect or modification;*
while, if stars of enormous comparative dimensions, were scattered through space, the likelihood is, that some one such body would be
sufficiently near us *to permit of our recognising
it under less ambiguous characters.*

Many other appearances—admitting of no
plausible solution on the supposition that all
those dim lights are sent from remote and ac-

Plate XII

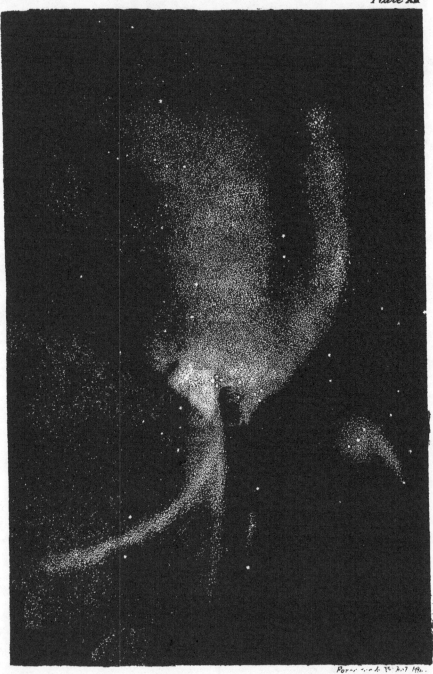

cumulated stars—sustain the inferences just
deduced, and thus greatly augment their probability. The wonderful Nebula in Orion, is in
this respect a most instructive phenomenon.
On directing the unaided eye to the middle part
of the sword in that beautiful constellation, the
spectator fancies on the first impulse that he
sees a small star; but closer observation shows
him that it is something indefinite—hazy—
having none of the distinctness of the minute
stars. When he looks at the spot through a
small telescope, these suspicions are confirmed;
and as the power of the telescope is increased, the more diffuse and strange the object.
Its form, as revealed by a twenty-feet reflector, is shown in Plate XII.; but there is reason
to believe that its low situation in our latitudes,
conceals many remarkable radiating branches
which are seen in the southern hemisphere.
Now, observe two facts;—the Nebula is *visible
to the naked eye,* and distinctly visible through
glasses of small powers; and *the whole light and*

efficacy of the forty-feet telescope could not resolve it into distinct stars. But, to be irresolvable by the largest telescope, the stars in the Nebula— supposing it a cluster—must be placed at a distance from us, which we cannot express in language; and to enable them to send us even a milky light through so vast an interval, they would require *a most improbable compression,*— improbable because unknown in degree even, in any explored portion of the universe. The hypothesis of a filmy or Nebulous fluid shining of itself, is thus again forced upon us, precisely as in the case of the Nebulous stars; and our general argument is here farther and very strikingly supported by the ascertained peculiarities of the mass. When telescopes are not sufficiently powerful to resolve a cluster, it still commonly takes on a succession of appearances, which distinctly indicate to the experienced observer—its resolvability, or stellar constitution. In the Nebula in Orion however, no such change appears. It grows brighter in one

sense, the larger the telescope, but only to become more mysterious. As we then see it, the
illumination is extremely unequal and irregular.
" I know not," says Sir John Herschel, " how
to describe it better, than by comparing it to a
curdling liquid, or a surface strewed over with
flocks of wool—or to the breaking up of a
mackarel sky, when the clouds of which it consists begin to assume a cirrous appearance. It
is not very unlike the mottling of the sun's
disc, only—if I may so express myself—the
grain is much coarser, and the intervals darker;
and the flocculi instead of being generally round
are drawn into little wisps. They present, however, no appearance of being composed of stars,
and their aspect is altogether different from that
of resolvable Nebulæ. In the latter we fancy
by glimpses that we see stars, or that could we
strain our sight a little more, we would see
them. But the former suggests no idea of
stars, but rather of something quite distinct from
them."—This great Nebula seems to occupy in

depth the vast interval between stars of the
second or third, and others of the seventh or
eighth magnitudes, and its superficial extent is
probably corresponding.   Its absolute size is
thus utterly inconceivable; for the space filled
by a Nebula of only 10′ in diameter, at the
distance of a star of the eighth magnitude,
would exceed the vast dimensions of our sun,
at least 2,208,600,000,000,000,000 times!

Although to the interruption of our course of
logical proof, I cannot refrain from adverting to
some of the engrossing contemplations, which
never fail to occupy me, when I gaze upon this
remarkable substance.   What is the intention
of such a mass?   Is it to abide for ever in that
chaotic condition—void, formless, and diffuse
in the midst of order and organization,—or is it
the germ of more exalted Being—the rudiments
of something only yet being arranged?   Then
too—although these questions were answered—
what is its present state?   It is not enough to
tell us that for such and such ultimate purposes

a certain object is destined,—we would know farther, the peculiarities and adaptations of its present or actual constitution? No part of creation exists merely as a *means*;—every thing is besides an *end* to itself; and within that looming mass, whatever be its final destiny, there are doubtless wide and systematic relationships, —each particle of its matter will be arranged and adjusted to its neighbour; nay, who can tell, who that has looked on those monuments of bygone worlds—the fossil relics which mark the early progress of our own planet—but, this amorphous substance may bear within it, laid up in its dark bosom—the germs, the elements of that LIFE, which in coming ages will bud and blossom, and effloresce, into manifold and growing forms, until it becomes fit harbourage and nourishment to every varying degree of intelligence, and every shade of moral sensibility and greatness!

Probable evidence of the foregoing nature

might now be almost indefinitely accumulated.
For instance, the magnificent appearance in the
girdle of Andromeda affords similar conclusions.
This Nebula, represented by Fig. 1, Plate XIII.,
is distinctly visible to the naked eye, seeming
like a greasy spot upon the dark blue of the firmament, or a light shining through a horn;
but as with the Nebula in Orion, *no telescopic
power has yet sufficed to give it the resolvable
aspect.* Farther pursuit of such considerations, however, is unnecessary, inasmuch as we
are in a condition to produce, what — taken
in supplement—amounts to positive and direct
proof at once of the reality and extensive diffusion of the Nebulous substance. I request your
attention to the phenomenon of COMETS.—There
is much connected with the comets of which we
yet know nothing; but two general and essentially characteristic facts are established, and
these suffice for my immediate purpose. In the
first place, the phenomenon demonstrates not
the *possibility* merely, but the actual *presence in*

Plate XII

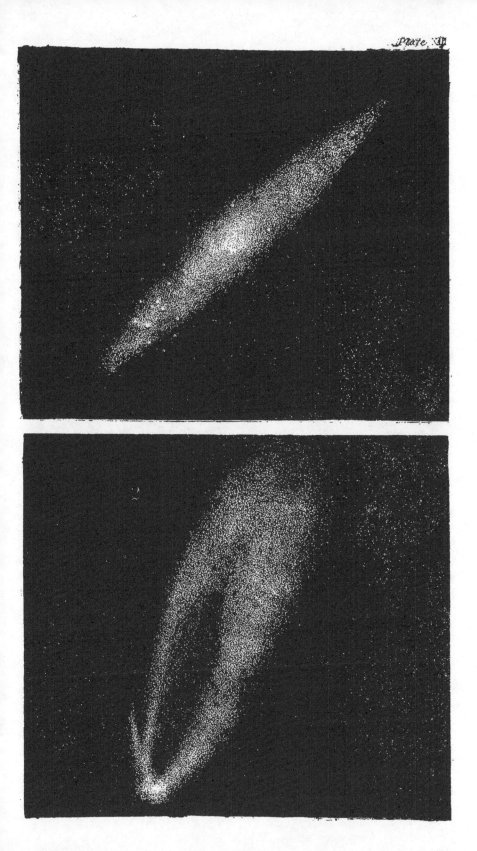

Nature of a nebulous modification of matter. The comets are nothing but *nebulosities*, small portions of a substance precisely similar in physical constitution to that which our hypothesis assumes. Even their nuclei dissolve into a fog under the inspection of the telescope. Fig. 2, Plate XIII. is a sketch by Sir John Herschel, of the second comet of 1825, and through the heart of another, the same observer once descried a cluster of stars of the sixteenth magnitude. Secondly: These small nebulosities are not connected with the structure of our solar system; from which we infer that *they are connected with some system in the spaces external to our limited sphere.* There is no essential tie between us and these comets; the variety of directions from which they come, altogether distinguishes them from the bodies which roll around the sun with singular and systematic regularity; they are chance visitants, most of them perhaps never approaching us but once, for, unless in a few instances, there is little

I

reason to believe that their eccentric paths are
continuous, or that they re-enter into themselves,
and form a definite and bounded curve. But
shall we therefore go into the usual inference,
that the comets are mere anomalies—freaks of
nature? Because they have no connexion with
the order of our planetary worlds, is it necessary
that they should have no meaning—no place in
the universe? Look around you! What is there,
what existing creature,—which has not such a
place? Of the fine web of Being, fitness and relation are the warp and woof. Apparent anomalies are mere finger-posts, pointing where things
lie of which we continue ignorant; and when
such intimation is received with philosophic
meekness, it invariably guides to unexpected
discovery. These hazy bodies, now and then
reaching our system, and leaving it without ever
operating an appreciable effect, are *not* spectral
and isolated monstra! As all things have a home
in nature, they too doubtless hold relations with

some grand external scheme of matter in a state of similar modification: and since, when influenced by the sun's attraction, they approach us from all quarters of the heavens, *the nebulosities in which they have their* ROOT, *must lie around us on* EVERY SIDE, *and be profusely scattered among the intervals of the stars.*   What an error to fancy these Comets anomalies! They demonstrate *that,* which, as we have seen, is required to make a large and varied series of phenomena explicable.   They are, in fact, absolutely indispensable; for without them the conjectural disclosures of the telescope would scarcely be established.   And in accomplishing this service, they have also vindicated their own position; so that we have at once two of our best intimations that knowledge is advancing,—remote phenomena appear in closest relationship, and objects and occurrences formerly deemed insignificant, assume a place as constituents of the compact fabric of the Universe.

We touch on the most obscure problem of Astronomy. As every atom in existence has its object, this nebulous matter, found in such abundance, must have a prominence in purport answering to its prominence in magnitude. But when we ask, *What are the nebulæ?* we feel that we are venturing into that dim twilight which always surrounds the sphere of positive knowledge. If we would understand them, however, or know whether they are intelligible, we must examine if they can be arranged under characteristics or peculiarities of *structure*, indicative of the operation of LAW; and it was the endeavour so to arrange them which led the Astronomer whose torch has hitherto guided us, to conjectures promising, now more confidently than ever, to throw unwonted light upon the course of material transitions. I have by means of diagrams, illustrated this part of our subject so fully, that I am encouraged to hope my reader will easily follow the arguments we are about to entertain.

Plate XIV

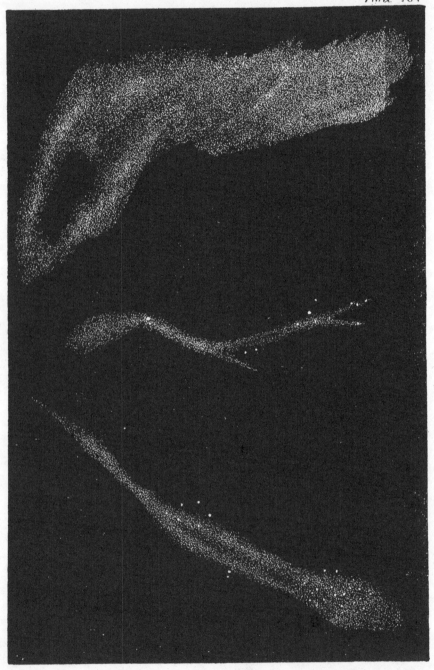

Plate XV.

Forrester & Nichol lithog.

In its first or rudest state, the Nebulous matter is characterised by a *great diffusion.* The Nebula in Orion is an example of this, and Plate XIV. represents a few other instances in which the milky light is spread over a large space so equably, that scarcely any *peculiarity* of constitution or arrangement can be perceived. The perfectly chaotic modification here illustrated, is perhaps the nearest to the original state of this matter, of any thing now remaining in the firmament; but the reason of its being found in separate patches, varying so much in form, manifestly appertains to remoter inquiries and an inaccessible period in the History of Things.

Parting from these perfectly diffused and amorphous Nebulosities, *Structure* as governed by *Law* begins to appear.   Even its first visible indications are very emphatic. The winding Nebulosity in Plate XV. for instance, exhibits a congregating or condensing of the filmy matter in two distinct places, which look like bright

nuclei, surrounded by a comparatively dark
ring, precisely as if they had been formed by
an actual *condensation* of the diffused matter,
under control of the law of universal attraction.
This is no anomalous appearance, for, *in every
case*, the seeming commencements of structure
are of the same kind. Nothing is any where
met with, like a dispersion, or indication of a
dispersive power; which probably would have
been seen, had any power but an aggregating or
condensing one, been influential over the condition of nebulosities. This aggregating power,
indeed, without the interference of any other,
appears to lead to the entire breaking up of
amorphous masses. The number of nuclei
which are found in distinct nebulæ is variable;
but there is never a departure from the character
due to their supposed origination in a condensing principle. We have always a regular gradation in their intensities. The point of light
grows brighter, while the vacancies or dark
circles around the points are marked with more

Plate XVI

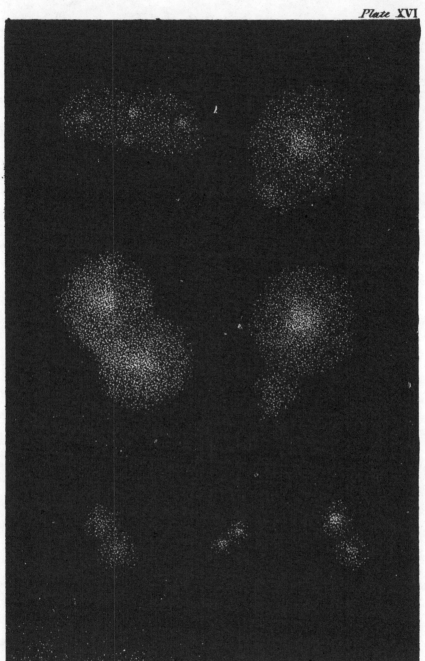

and more distinctness, until we reach a cluster
of small round nebulæ perfectly detached from
each other.   The first figure in Plate XVI. represents an early stage of the phenomenon described; and the others in the same plate carry
on the process of apparent separation in regard
of two nuclei, until the masses are altogether
distinct.   The progress may be traced still
farther.   In their ultimate condition—that condition I mean in which they seem to merge into
stars—we find them distinguished from a double
star or a cluster, only in this, that they rest on
a bed of very faint light.   Of course when I
speak of *progress*, I will be understood to sig-
nify the progress through a series of related
contemporaneous objects in different and gra-
duated states—not that progress which has
never been observed—the passing, viz., of one
nebula from an inferior into a higher condition.
But is the conclusion rash, that this latter progress is possible?   Is it not darkly intimated
by the unbroken integrity of the series; and

may not the great law which explains these
structures—so various, yet so related—have actually brought such distinct round nebulæ, as
are seen clustering together, and also collections
of stars imbedded near each other on a whitish
light, from the bosom of masses like those looming elsewhere—still all indefinitely—among the
recesses of our firmament?

The hypothesis I have started is so strange,
and brings up notions so unlike our common
thoughts of the stellar universe, that it cannot
and ought not to be received without an almost
superfluity of evidence. We turn, accordingly,
to single and definite Nebulæ which are possessed of marked structure, in search of what
light they may chance to throw upon it.   And
it happens that we get nothing here but con-
firmation. Their shapes, and the distribution
of light in these separate bodies so entirely
accord with the hypothesis of condensation,
that we have hardly room for escape.

Plate XVII

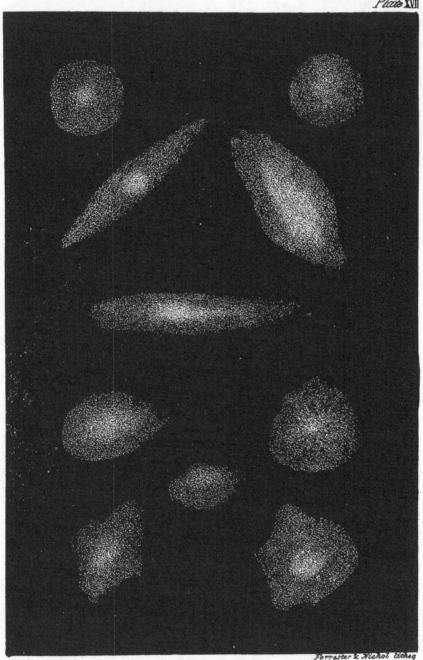

Forrester & Nichol lithog

Single Nebulæ characterised by structure, vary in form,—the usual varieties are shown in Plate XVII. Some are of an oblong shape, like that in Andromeda, with marked condensation towards the centre. We know not how these flat zones originated, but even in them structure is always manifested in a form indicating condensation around the apparent centre of the mass. The great proportion of distinct nebulæ, however, are round, or nearly so. In Sir John Herschel's recent catalogue the immense proportion are marked as round. It is a fair, and almost necessary inference, that these round masses are *spheres;* for there is no likelihood, that if they were cones or cylinders, they would all so lie in regard of our line of sight as invariably to present their circular sections. Now the sphere is the shape *naturally assumed by masses whose particles mutually attract each other;* and on the principle that the globular shape of the rain-drop is accepted as a common illustration of the all-prevalence

of gravity, and the sphericity of the planets held
to confirm the same important law, this tendency in the nebulæ to the round form must be
received as a weighty and important argument.
But it is the mode of the distribution of their
LIGHT, which most confirms and settles these
views. On the very first glance, this seems
a phenomenon *connected with central influence ;*
for not only is the illumination uniformly greatest at the centre, but if in any nebula a circle of
any radius be described around the centre, the
illumination will be found *of the same intensity
at every point of its circumference.* By far the
most remarkable and important fact however is,
that wonderful *gradation* in the intensity of the
central light. Plate XVIII. speaks to the eye,
and is more valuable than pages of description.
Each figure in that plate is the representative,
not of an individual, but of an extensive class ;
and surely a series so well marked—so striking
in its aspects—must indicate the presence and
influence of a great law ! From absolute

Plate XVIII

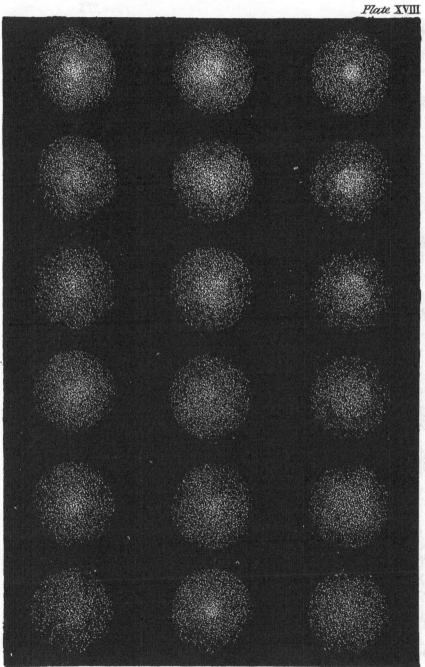

Plate XIX

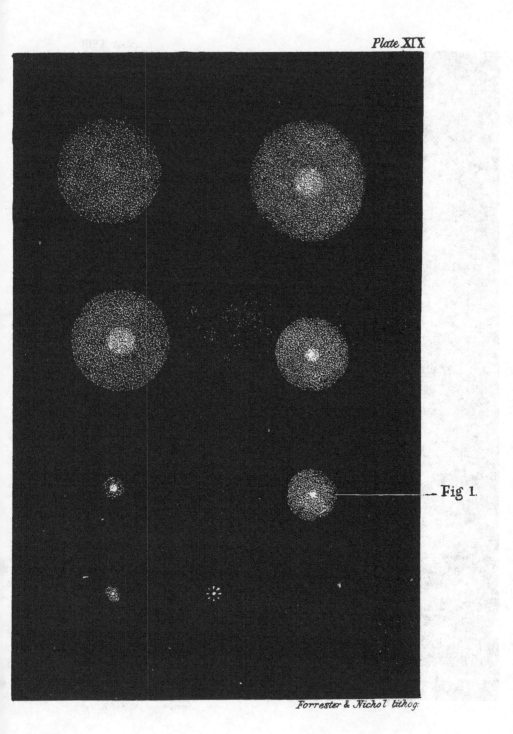

Fig 1.

Forrester & Nichol lithog:

vagueness, to distinct structure, and then on to
the formation of a defined central nucleus, the
nebula seems growing under our eye! The
illustration of Laplace, reproduced by Mr Airy,
here forcibly occurs to me : " We look among
these objects as among the trees of a forest ;
their change in the duration of a glance is indis-
coverable ; yet we perceive that there are plants
in all different stages ; we see that these stages
are probably related to each other in the order
of time, and we are irresistibly led to the conclusion that the vegetable world in the one case,
and the sidereal world in the other, exhibit at
one instant a succession of changes requiring
time, which the life of man or the duration of
a solar system may not be sufficient to trace out
in individual instances." And the progression
still goes on. We have objects such as the
Figures in Plate XIX. ; until in the end a STAR
is found perfectly organized with a mere *bur*
around it !

The circumstance which bestows so much verisimilitude on this whole speculation, and which most of all induces us to think that the surprising series we have just reviewed really indicates the progress of worlds from chaos, is our previous knowledge of the wide operation of a law capable of converting one of these inferior nebulæ into a member of a more perfect class. We institute no new power—we do not speculate on the *possibilities* of Nature; that Nature is uniform—that GRAVITY which controls the planetary, and, in so far as we know, all celestial spheres, acts also within these masses, is our boldest hypothesis.   And there are positive indications that it does act there.   The comets are under its influence, for they are attracted by our sun : their matter, therefore, is the same as planetary matter—obedient to gravity.   Instances, too, are not wanting in which the original distribution of external nebulosities appears affected, or altered by the attraction of neighbouring stars.   The upper and lower figures in

Plate XX.

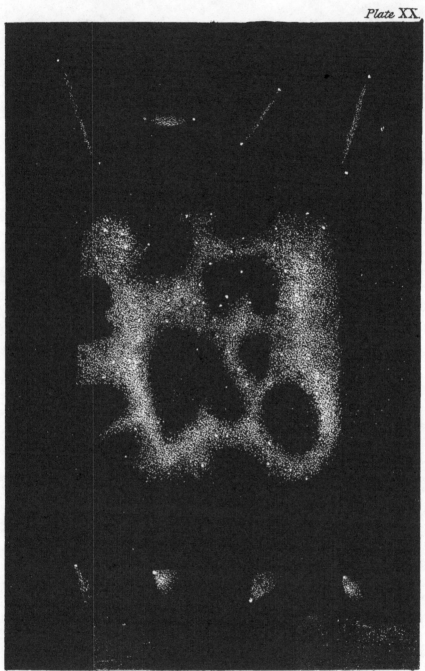

Forrester & Nichol lithog

Plate XX. show portions of it being drawn towards a star—as if feeding it—or stretched between two. The middle object in the same Plate is very instructive. It is a diffused nebulosity near a reticulated cluster of small stars, and its brightness follows their line; while in the spaces where no stars are, there is no nebulous matter, as if it had been drawn away by attraction, and accumulated in those bright ridges, or absorbed.

It may indeed be said, " Show us a change, show us the actual progression of one nebula from an inferior to a more organized condition. We grant the unbroken completeness of your series; we grant the law which could advance its individual constituents ; but *actual change* alone is proof." Let us not, however, be deceived, or mistake the nature of the problem which is engaging us. If the nebular hypothesis be true, all the forces developed upon the thin surface of our planet, and which have given rise to geological transitions, stretching through

periods in which the existence of the human
race is an invisible speck, will have resulted
during a stage of condensation in a secondary
nebula, which no instrument from any fixed
star could possibly detect. How then delude
or disappoint ourselves by straining the eye after
periods so enormous! There is a creature named
the Ephemeron, whose life is confined to the
veriest point of time,—in one short hour it
dances out its existence in the sunbeam. That
creature is in presence of all the phenomena of
vegetable growth; it may see trees—it may
see flowers, but how could it or its generations
actually observe their progressive developement?
In relation to the nebulæ Man is only an Ephemeron. Fifty lives succeeding each other, and
of a length to which individuals often attain,
would reach backwards beyond the recorded commencements of his race; and in the mutability
of things, fifty more may constitute a line longer
than his allotted epoch. And, no more than
one hundred of those creatures, which are born,

breathe, and die, could learn of the progress upwards of the majestic pine,—will man ever learn
of the changes of the nebulæ!

The ideas I have now presented to you—
august and strange though they are—should not
appear in contradistinction to what every moment is passing around us. Supposing these phenomena did unfold the long growth of worlds,
where is the intrinsic difference between that
growth and the progress of the humblest leaf,
from its seed to its intricate and most beautiful
organization? The thought that one grand and
single law of attraction operating upon diffused
matter may have produced all those stars which
gild the heavens, and, in fact, that the spangling
material universe is, as we see it, nothing other
than one phase of a mighty progress, is indeed
duly surprising ; but I appeal to you again in
what essential it were different from the growth
of the evanescent plant? There, too, rude matter puts on new forms, in outward shape most
beauteous, and in mechanism most admirable :

and there CANNOT be a more astonishing process or a mightier power even in the growth of a world! The thing which bewilders us is not any intrinsic difficulty or disparity, but a consideration springing from our own fleeting condition. We are not rendered incredulous by the *nature* but overwhelmed by the *magnitude* of the works; our minds will not stretch out to embrace the periods of this stupendous change.   But Time, as we conceive it, has nothing to do with the question—we are speaking of the operations and tracing the footsteps of One who is above all time—we are speaking of the energies of that Almighty Mind, with regard to whose infinite capacity a day is as a thousand years, and the lifetime of the entire Human Race but as the moment which dies with the tick of the clock that marks it—which is heard and passes.

LETTER VII.

THE NEBULAR HYPOTHESIS.

LET us note the exact amount of evidence constituted by the speculations of the foregoing letter, on behalf of the Hypothesis that all existing stellar bodies sprung by virtue of the law of attraction, from the bosom of a chaos similar to the vague masses I have described. In so far as this Hypothesis undertakes to EXPLAIN THE NEBULÆ, I do not conceive that much of *accessible* knowledge is now wanting to confirm it; for, the agreement of the forms of the nebular substance with the natural results of the persevering action of gravity, seems almost demonstrated. But it must not be forgotten that there is another correlative and very exten-

K

sive inquiry which this truth has not touched;
The Hypothesis must also EXPLAIN THE STARS.
If it is the true Cosmogony, and we have at
length approached a right theory of the Formation of Things, we should indeed obtain from
it a satisfactory idea of the meaning of that
curious progression of structure, which so strikingly characterises the Nebulous masses; but it
is no less imperative that it exhibit with proper
distinctness, how the mass of stars around us,
along with their peculiar features and arrangements, might have been evolved, in obedience
to *known mechanical laws,* by the condensation
of Nebulæ.   To the inquiry thus suggested, I
invite you rejoicingly; at every step in our pursuit of it, we will gain new views of the unity of
Things, indications of remote and unexpected
relationships, and proofs the most illustrious
afforded by Science, of the compactness of that
Domain whose forms occupy SPACE, and the
annals of whose changes constitute TIME.

I. There is indeed a difficulty in reconciling the imagination to the idea that an orb like our SUN—on which, as in our former survey, we naturally first cast our eyes—could have originated in a vague nebulous mass. Observation shows, however, that the *magnitude* of our luminary is no obstacle to the Hypothesis, for the statement in page 126 proves that a Nebula like that in Orion, contains matter or substance sufficient for the generation of a solid globe perhaps some millions of times as large. Neither can the difference between the *solidity* of the Sun and the *gaseous* condition of the Nebula, constitute ground for rational hesitation, inasmuch as in the laboratory of the chemist matter easily passes through all conditions, the solid, liquid, and gaseous, as if in a sort of phantasmagoria, and his highest discoveries even now are pointing to the conclusion, that the bodies which make up the solid portion of our Earth may, simply by the dissolution of existing combinations, be ultimately resolved into a per-

manently gaseous form.*   But it should wholly
reconcile you to the preliminary conception, if,
partly by what seems the true interpretation of
a phenomenon long noticed, and partly by aid
of one of the most promising discoveries of
modern times, I show you that the SUN is not
yet pure, that he *has not yet quite escaped that
original nebulous character* I am attributing to
him, and that, notwithstanding of his effulgence, he is still rather in the condition of a
nebulous body—something like Fig. 1, Plate
XIX., where consolidation although very far
advanced, is not perfect.

The first appearance I allude to is the
ZODIACAL LIGHT.   This remarkable phenomenon consists in a long train of faint light
of a conical form left by the sun after setting and projected on the sky.   It is easily
seen in tropical countries, but in ours it is
only visible when in the most favourable posi-

---

* See Note C, at close of the volume.

tion; it may be looked for on clear evenings in March or April. It has an appearance like that in the subjoined wood-cut, where S is the sun.

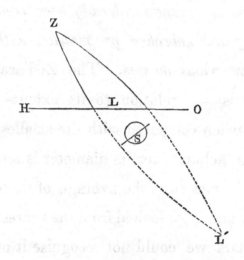

H O the horizon, and Z L the faint light. Now, the axis on which the sun rotates is, as shown in the Figure, at *right angles to the greatest length or axis of the light;* in which alone, we trace a connexion between that light and the axis or equator of the sun. Although we see only the upper part of it, it is thus perhaps extended also on the sun's other side as in the dotted part of the line; forming a regular and still uncondensed relic of some *oblong nebula,* in which our luminary probably

originated. We are at least certain, that this
light is a phenomenon precisely similar in kind
to the nebulous atmospheres of the distant stars ;
and, *these atmospheres have already been connected by a long and unbroken progression with
their parent amorphous masses.* The Zodiacal
Light, indeed, is—in relation to its extent—
utterly trifling when compared with the smallest
of these stellar nebulæ; for its diameter is less
by millions of times than the average of these
curious bodies, and if we looked from the nearest
of the fixed stars we could not recognise it or
guess its existence with the aid of Herschel's
largest telescope : but this fact, so far from being
barren or forbidding, is itself a ground of important and cheering inference. Previous even
to the discovery of such a circumstance, no one
could have said with confidence that the nebulous
atmospheres of the stars terminated actually at
the point of magnitude below which we cannot
see them, or at the varying limits of their visibility to us ; but we are now furnished with
substantive reason to believe that the pheno-

menon of the illuminated bur *does* extend much farther—that vast numbers, if not the whole, of the bright and apparently pure constituents of our firmament, may still have Zodiacal Lights; and that the nebulous phenomenon, although to our imperfect vision now characterising but comparatively few objects, may, notwithstanding of the long lapse of hither time, continue to possess a generality of which all that we see is only the faintest indication !

The other and new phenomenon—one no less remarkable in the mode of its discovery, than because of its intrinsic value—is equally confirmatory of the nebulous attributes of the sun. The nebulous stars, as seen by our telescopes, have very commonly a well marked increase of light towards their centre; or, what is the same thing, a gradual diminution, or shading off, towards their edges. Now this Zodiacal Light is probably only the *brighter part* of the solar nebulosity, which may extend much farther. In this further extension

it may not give out sufficient light to be perceptible ; and the fact is, if it includes the Earth,
we could not see it, although it emitted a very
considerable quantity of light. Being in the
midst of it, its light could evidently not be
observed as an *external* thing : it would be con-
founded with that faint illumination arising from
the intermingling of the rays of all the stars,
which somewhat brightens the dark ground of
our skies ; and must have remained unknown,
but for another mode of experiment. In note A,
at the close of the volume, where the planetary
phenomena are rapidly surveyed, I have shown
how the permanence of the orbit of each planet
depends upon the perfect balance of two forces
or tendencies, viz., the attractive power of the
sun, and that tendency to fly from the centre,
which follows from the motion of bodies being
naturally in straight lines, and whose energy
depends, in each case, upon the rapidity of the
body's motion. If the power of either of those
balanced forces be diminished, it is clear that

the authority of the other will prevail.   Relax
Gravity therefore, and the planet will recede
from the sun, and its orbit widen until a balance
is restored.   In the same manner, diminish the
rapidity of the body's motion, and, as the centrifugal force will be diminished by the same act,
Gravity will prevail ;—so that the body's orbit
will be *contracted* or *drawn in*.   Now if a nebulous fluid is diffused through the planetary
spaces, every body which moves through it must
experience resistance and be *retarded* as we are
by the atmosphere when riding at a rapid pace ;
and we would thus expect to trace the ether's
existence in the fact of the planetary orbits gradually drawing in, and the revolving bodies approaching the sun.   Unhappily, however, for this
only mode of observation left, the planets are too
dense, too large to be of service in so delicate an
inquiry.   However light and thin the ether,
there is no doubt that it must and will influence
even *their* motions ; but perhaps, by a quantity so small, that the accumulation of the per-

turbations arising from it during the whole
period of accurate astronomy, would not render it perceptible. No trace of such influence,
indeed, is yet found in our planetary tables; and
astronomers would have been left in regard of
the whole subject to conjectures, which however plausible had yet no actual or experimental
ground, unless for a remarkable and certainly an
unlooked-for occurrence. Until recently, astronomical science has not been able to present a
complete and minutely accurate view of the orbit
of any comet. The general character of the
orbits of these bodies, and the important elements, at least of *one* of them, have been known
since the time of the celebrated Halley; but this
philosopher knew nothing further than the general elements ; and no orbit was laid down with
exactness sufficient for the above purpose, until
Encke of Berlin examined with so much accuracy the conditions of a *body*—if a thing so small
and vaporous merits the appellation—which
completes its eccentric course around the sun in

$3\frac{1}{2}$ years. Now, it appears by observation, that this comet *is approaching the sun ;*—on every successive appearance, we find its orbit some-what contracted, and there is reason to believe that the contraction will go on until it is either absorbed in that luminary, or altogether dissipated by his beams. And after vainly searching for some other cause, enquirers are nearly una nimous in referring this extraordinary and hiherto unparalleled change, to a RESISTING MEDIUM or ETHER occupying the planetary spaces. " I cannot but express my belief," says Professor Airy, " that the principal part of the theory, viz., an effect exactly similar to that which a resisting medium would produce, is perfectly established by the reasoning in Encke's memoir ;" and similar opinions have been offered by other great authorities. That the sun, then, has a widely diffused nebulous atmosphere —extending far beyond the limits of the zodiacal light, and if not beyond, at least deep into the planetary spaces—an atmosphere of which that

light may merely be the densest portion, is at
length rested on a high degree of probability :
and how singular is it that we should have been
guided to a truth so remote and difficult—one
concerning which the grandest phenomena of
our system are silent, by the motions of a wan-
dering object, in comparison with whose ethereal
nature, even one of these light flocculi or flakes
of cloud, which scarce stain the sky of a summer
evening, is perhaps heavy and substantial!
Even thus harmonious is the universe! And
seeing the perfect continuity of that golden line
of order, which unites its mightiest phenomena
with its least, so that the motions of a speck of
dust may illustrate causes adequate to generate
worlds,—what achievement is too high to be
hoped for future discovery, and why may not a
time arrive, when in return for man's close observation, and unwearied questioning of nature,
the darkest of those speculations in which we
have just been indulging, shall be doubtful and
venturous no more?

If consideration of these circumstances has sufficed to destroy your natural repugnance to the resolution of our effulgent sun into a Nebula, you will be prepared to start the next question—the question as to the capacity of the Hypothesis that the sun sprung from such a mass, to explain the remarkable feature of his rotation around an axis.

When we reflect on the Solar nebula in the act of condensing, it appears that the act consists in a flow or rush of the nebulous matter from all sides towards a central region; which is virtually equivalent, in a mechanical point of view, to what we witness so frequently, both on a small and large scale—the meeting and intermingling of opposite gentle currents of water. Now what do we find on occasion of such a meeting? Herschel's keen glance lighted at once on this simple phenomenon, and drew from it the secret of one of the most fertile processes of Nature! In almost no case do streams meet and intermingle, without occasioning where they intermingle, a *dimple* or WHIRLPOOL; and in fact, it is barely possible that such a flow of matter from opposite sides could be so nicely balanced in any case, that the opposite momenta or floods would neutralize each other, and produce a condition of *central rest*. In this circumstance then—in the whirlpool to be expected where the Nebulous floods meet—is the obscure and simple germ of rotatory movement. The very act of the condensation of the gaseous matter as its flows towards a central district, almost necessitates the commencement of a process, which though slow and vague at first, has, it will be found, the inherent power of reaching a perfect and definite condition, and from which consequences ultimately issue, not less various and astonishing than the foliage and stature of the noble tree, considered as the developement of an insignificant seed.

You will have no difficulty in perceiving, that —the whirlpool motion once originated—there

is an inherent power in matter, under such circumstances, to evolve finally a definite rotation
of considerable velocity. It is a general law or
fact, that if a body is subjected to two moving
influences acting in different directions, *it obeys
both*, or moves in obedience to each, as if the
other did not exist. For instance, let a person in

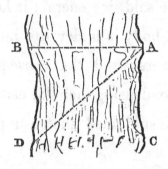

a boat start from A, with the intention of pushing his boat right across a stream; and suppose, that in the time he would occupy in rowing
from A to B, *if the water were still*, the power
of the stream could carry him from A to C *if
he did not row at all*;—the question is, how his
boat will actually move? Now, all experience
tell us, that if he merely rows *right across*—
pushing his boat from the bank A towards B,

with its side invariably to the stream, he will,
previous to reaching the opposite bank, have
been carried down by the current, precisely as
far as if his boat had merely floated; so that he
will reach the bank at D, immediately opposite
C, and his boat will have *partaken at once of*
the two motions or influences impressed upon it.
This law, as I have said, is general; it holds in
every case where a body is under the influence
of two moving powers: and the consequences
of its action in a condensing nebula, cannot be
mistaken.   Let the subjoined sketch represent

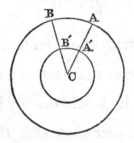

a section of a circular nebula, revolving about
the central region C, and in which, condensation is permanently going on. It is evident that
the particle at A, in consequence of the whirlpool, moves from A to B, while the particle at

A', only moves from A' to B'; but, as the attractive power, by drawing the first particle from A to A', cannot by the foregoing principle, *diminish its circular velocity*, the result of such condensation will be *the attaching to* A' of another particle A, whose circular velocity is greater than its own. Now, the permanent consequence is manifest. If two balls, A and B for instance, are moving forward with different velocities, A much faster than B, what

will follow when A overtakes B ? Certainly an acceleration of B's motion, and a retardation of A's ; and the two together will, after contact, move on *much more rapidly than at B's former rate:* so that, by the very act of A (see Fig. on previous page) being brought into union with A', the rotatory velocity of A' would be augmented; and if the whole *outer* circle A B, &c., were attracted towards the inner circle

L

of matter A' B', &c., that inner circle would accordingly rotate more rapidly than before, and the velocity of the rotation of the entire nebula, must therefore be increased. Physical objections, I am aware, may be taken to this explanation ;—I propose it merely as a *popular* one : but it indicates, nevertheless, the principle which assures us that the condensation of a diffused and comparatively slow whirlpool, cannot take place without the great and growing increase of the velocity of its rotation, inasmuch as the momentum or amount of the *rotatory force* must, in all its stages and conditions, continue the same. And thus, if you have followed me, you see how out of phenomena the rudest and most unpromising, and by the simplest laws of nature—those which guide the facts of every-day experience,—even that stupendous rotation might be generated—a rotation whose discovery was one of the first achievements of the telescope, and which, all who know Nature, ought to be assured, does not stand by itself or as an

independent fact, but is a cosmical phenomenon
of wide significancy, and closely however mys-
teriously related with the whole scheme and
progress of Things.*

II. Our conclusions are already of great importance, and will lead us far. Not only
are we now able to look upon these myriads
of single suns, as having come—as probably
at least as our own luminary—from the womb
of the Nebulæ; but we have likewise generalized the phenomenon of rotation by gene-
ralizing its cause, inasmuch as the foregoing

---

* The velocity of the resulting rotation will manifestly depend on the magnitude and condition of the original Nebula;
so that we would expect to find no uniformity of period in the
rotations of the stars—an expectation, perhaps verified in the
phenomena of the *variable* orbs. The probable variations of
size in the original Nebulæ, also leads us to the supposition
that there is probably a very great variety in the magnitude
of the resulting stars—a supposition appearing to invalidate
our first Hypothesis, as stated in Note at page 41. Be it
remembered however, that this admission does not destroy
our conclusions; it merely demands their modification. Our
object was, not to establish or communicate perfectly accurate
formal truths, but to learn somewhat of the structure and
extent of the Universe.

demonstration applies to any condensing mass :
and from the vantage ground of this farther
and almost unexpected insight into phenomena, we can proceed with some confidence
to take cognizance of the general fact next
in order of simplicity — viz. that evolved by
Herschel's fine induction from the existing
statistics of the Heavens — the law of the
*revolving motions of stars associated in limit-
ed and compressed clusters.*   The solution of
this great and interesting law afforded by our
Hypothesis, I cannot term less than pictu-
resque,—it excites instantaneously our surprise
and admiration.   Have you ever walked in a
mood of tranquil thought along the side of a
quiet river, whose waving banks reflect a thousand currents, by the intermingling of which
numerous dimples or whirlpools are produced—
their easy glide only marking the river's still-
ness ?   Have you seen these dimples follow
and pursue each other as if in gambol, or
watched the phenomenon of the near approach
of two or three ?   Then have you witnessed

the secret of the mystery of the double and triple stars! When one of these dimples reaches the verge of another, they begin *to revolve around each other;* and in fact they must, on approximation, act upon each other as *two wheels,* so that a revolution of each around the other *must* immediately supervene, and increase in rapidity, until by external pressure, they are forced into one.   Plate XVI., in which Double Nebulæ are presented in various stages, enables us to apply this illustration. If the single nuclei are rotating, as we are now almost entitled to say they must be, it is precisely a case of two contiguous whirlpools, and *how could revolutionary motion be prevented?*   Two such masses in approximate contact *must* originate such a motion: as the principle of gravity draws the nuclei nearer each other, the velocity of revolution will manifestly increase; and the two bodies will constitute themselves into a stable system when the rapidity of revolution suffices to counterbalance their mutual attraction.   The case is manifestly the same in instances of three, four, or more nuclei, formed in the immediate neighbourhood of each other, or out of one such mass as Fig. 1, in the same plate; so that we have now a CAUSAL solution of Herschel's remarkable prophecy. This part of our subject, too, happily presents a point of verification.   All known double Nebulæ should be sedulously watched; although they may move slowly, still their motions might in some instances be detected,—a positive discovery which I do not hesitate to allege would not be second in interest to that which has immortalized our age—the discovery of the actual motions of the double stars! I am not sure that the portion of the nebular speculation over which I have just gone, is not in my eyes the most engrossing of the whole of it, for it points emphatically to a moral I am extremely anxious to impress.   We are all too easily inclined to look on creation as made up of isolated parts—of independent or individual

classes of beings—and to regard Nature as
we do a case of botanical or mineralogical boxes;
so that it requires a fact as striking as the
identification of the Stellar motions of REVOLU-
TION with those of ROTATION, to startle us from
the habitual error, and to bring us to right
views of that stupendous ORDER within which
we live, and of which our own· beings constitute a part. The unity of things—their inter-
dependence—their adjusted relationships, are
proclaimed by every department of the Universe. I deny not that different laws may exist;
nay, they *must*,—for it is only by the com-
mingling of Opposites that Variety and Progress can be produced; but all is not opposi-
tion which seems so, and most of what we
divide and parcel out into isolated bundles, is
nothing other than the parts of the same
grand scheme. Philosophy has taught this for
ages—it is, in fact, the secret of her life; for
she aims to gather up all fragments and to pre-

sent the Universe united, compact, tending to one end—a type of its August CREATOR.

III. I am now to try the Nebular Hypothesis by our only remaining definite test—the most severe of all. As we come homeward, scenes and objects are better known, and even familiar; and although a plausible tale of foreign lands may pass off because of its general credibility, nothing less than minute exactness will be accepted here. Our SUN we know, is surrounded by PLANETS belonging to that star, and having manifestly originated in the process which brought it into an organized condition; and the question immediately presents itself, whether the origin of these small bodies, and of their peculiar motions and mechanism, is satisfactorily explained by the Nebular cosmogony ? If the discussions on which we enter are somewhat more difficult and tedious than the preceding, I believe their results will yet reward the exercise of your attention ; for we are about to connect the phenomenon of planetary existence likewise with the fundamental fact of the ROTATION of the condensing Nebulæ, and to demonstrate that law at which on several occasions I have hinted— the almost necessity of planets around every star in which the elementary principles of matter, as we know it, freely energize.

1. The preservation and permanence of the place of a revolving body depends, as I have said, on the circumstance that the centrifugal force is not greater than the power of the central attraction. The inevitable consequence of an excess of the former is seen in simple operation in a common phenomenon. You may have heard of a fact known to most Mechanics, that a grindstone may be made to revolve with a rapidity sufficient to cause splinters fly off from its rim, and even the whole rim to break in pieces—indicating that the centrifugal force of the rim with

that velocity, more than counterbalances the
mutual attraction or cohesion of the particles of
the stone. Now, if the rim, instead of being
formed of brittle stone, had consisted of an
elastic belt, say of caoutchouc, what would have
resulted in such a case? Clearly a separation
of the rim from the mass of the rotatory body—
it would have expanded somewhat, just as the
orbit of a planet in a similar position; and had
other circumstances permitted, would have revolved around the stone as a separate ring at a
distance where the balance or equilibrium of the
forces was restored. Plate XXI. will enable
us to apply these considerations to the case of
a condensing and therefore *rotating* Nebula with
striking effect. Fix your attention on Fig. 1.
We have already seen that causes continually
operate to increase the velocity of the Nebula's
rotation; but when this velocity became in any
case sufficiently great to communicate an over-
balanced centrifugal power to the exterior
portion, that phenomenon must result which

Plate XXI.

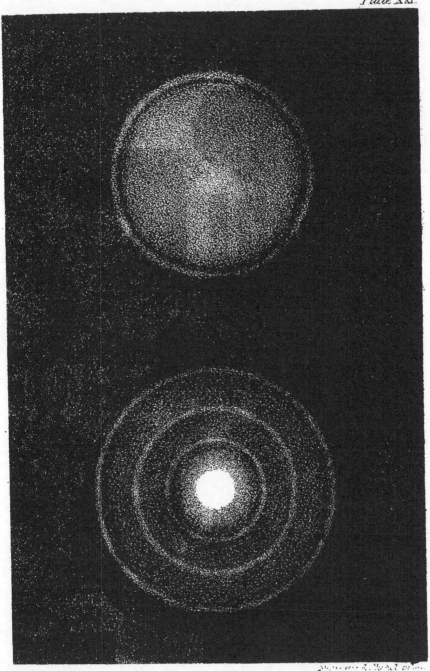

we have illustrated by the grindstone—the outward part of the elastic Nebulous fluid, would
somewhat widen its diameter, separate itself as
represented by the figure, from the parent mass,
and assume the position of a *distinct portion of
matter revolving in* SOME FORM *around the central body.* There is no doubt whatever of the
mechanical principles on which these inferences
rest; and it is equally certain that there are
almost infinite chances against the condensation
of any large or original Nebula, without the
occurrence of circumstances which would cause
it throw off numbers of such rings, so that, in
a more advanced condition, every such mass
might (*if the forms of the thrown off rings had
not altered*) present the appearance of the second
figure in Plate XXI.,—a large central nucleus,
with subservient rotating annuli, composed of
quantities of matter necessarily very small when
compared with the main body.

Here then, have we our first idea of the origination of planetary—or of quantities of subser-

vient revolving matter; and the next question
is what *forms* would these rings probably ultimately assume ?   There are three possible
forms.   1. The mass, if tolerably equable in its
original constitution, and undisturbed from
without, might condense as it is, or into a *rotating* SOLID RING ; but the chances against
such a result are so numerous that we would
expect the phenomenon to be very rare in the
Universe.   2. If the mass broke up or separated while condensing—as its own internal
irregularities would in all probability constrain
it to do—it might divide into a number of portions so equal in attractive energy that none of
them would have any tendency to coalesce with,
or fall into the others ; so that the ring would
ultimately be transformed into a number of distinct small solid bodies, revolving around the
central mass at nearly the same distance from
it.   These bodies, it is clear, would in their
final state be *spherical* or *round planets:* and
although not so evident, it is yet mechanically

certain, that they would *necessarily rotate on their axes in the direction of their revolutions.*
3. Even this second supposition, however, is not a very probable one, inasmuch as its essential condition—the attraction of the mass of the ring towards equally balanced centres—could in the nature of things occur but rarely. By far the likeliest result is the division of the annulus into nuclei of unequal power—the larger of which would, by its superior attraction, assume the others into its mass,—the whole solidifying into one considerable globe. Such globes would likewise invariably follow the law of rotation above specified; and every one of these secondary masses, might, during the phenomena of its subsequent condensation and augmenting velocity of rotation, throw off rings corresponding in all respects to the rings around the primary nucleus,—these condensing in their turn, and according to the foregoing laws, into solid annuli and *Satellites.*

How exact is the correspondence of these

general results with the character of the bodies
in our solar system! *First,* we have a central
massive globe, with subservient globes engirdling him at such distances, that when the
sun in a diffused state stretched out to their
respective limits, *his rotation must have been
equal to the periodic times observed by the planets
at the present day*—some small allowance being
made for changes which must have supervened
during the long lapse of ages. *Secondly,* The
fundamental principle of the theory being verified—in so far as it is capable of verification,
cast your eyes over the masses which compose
our luminary's cortége! 1. The great proportion
of the planets belong, as we would expect, to
the last of the three defined classes of forms into
which a ring might break up. MERCURY,
VENUS, the EARTH, MARS, JUPITER, SATURN,
URANUS, are single globes, revolving in orbits
of their own, and around some of them are dependent satellites. 2. In one instance only,
does the ring seem to have divided into equally

balanced parts—I allude to the four small planets, those ASTEROIDS between Mars and Jupiter, which have nearly a common orbit, or which revolve at almost the same distance from the sun : and 3. We have, also in one solitary instance, a specimen of that most singular of cosmical appearances—AN ORIGINAL RING, solidified in its pristine condition, and revolving around the planet SATURN.*

2. To have arrived thus easily at a connected origin of the several bodies composing our planetary scheme, is unquestionably a great achievement; but there still remains perhaps the most

* It is not unlikely that there are many other very small bodies which have come into being in the same manner as the others, revolving in various and *disturbed* orbits, within our planetary spaces. Meteors, or falling stars, are now generally believed to be most easily accounted for, by supposing them bodies of this sort drawn within the earth's atmosphere by attraction, and there ignited. The annual occurrence of a shower of meteors in the month of November, distinctly intimates the existence of an unexplained *astronomical* phenomenon ; because, by being connected with the ecliptic and a particular season of the year, it is connected with the position of the earth in its orbit.

difficult part of the problem;—the Cosmogony
must be reconciled with the condition of the *entire* system.   What I refer to, is as follows :—
The planets, without exception, move around
the sun in ovals differing little from circles, and
lying almost in the plane of the sun's equator;
they all revolve in these orbits in the direction
of the sun's rotation on his axis; they rotate on
their axes in the same direction, and—except-
ing what we have been told concerning the still
enigmatical and but partially known body Uranus—the whole Satellites, including the rings
of Saturn, revolve around the primary planets
also in that direction; nor are the rotations
of these secondary bodies, in so far as they are
known, subject to a different law.   Now, these
phenomena receive no explanation from what
we usually term the law of gravitation, inasmuch as gravitation could sustain systems distinguished by no such conditions,—nay, it actually does so, for the comets are free from all
these laws, they move in very eccentric orbits

inclined to the plane of the sun's equator at all
degrees, and their motions are as often retrograde as direct.   Most fortunate it was that an
inquiry baffling even the resolving power of
Gravity, and thus profounder than any under-
taken heretofore, fell into the hands of a philosopher whose knowledge of celestial mechanism
was then complete, and whose capacity to trace
elementary laws to their remotest consequence,
has never been surpassed!   It occurred at once
to the illustrious LAPLACE, that the ordinary
operation of Gravity is to sustain or regulate
systems which have been brought into being;
and that the higher conditions of which I have
spoken, pertained directly to the manner of our
system's origin ; nor did he meditate long ere
the splendid speculation I have detailed arose
in full maturity in his mind, and connected itself with the revelations men were at that time
first receiving from the telescopes of Herschel.
Observe how intimately the system of the generation from *rings,* co-ordinates with these con-

M

stituent phenomena. The rings in the first instance must be circular, and thrown off at the sun's equator where the velocity of rotation and therefore the centifrugal force is the greatest; the bodies resulting from them must revolve all in one direction, and with velocities corresponding to the velocity of the Nebula at the period of their separation; the separate and consolidated masses resulting from their destruction must as stated rotate on axes in the direction of their revolution; and finally, all Satellites subsequently formed must both revolve and rotate according to the same order! The Cosmogony has thus every mark of truth: its roots are *seen* in the Heavens, and it appears to go through every nook and alley of solar and planetary arrangements, not only explaining them but comprehending their variety and deducing the whole from one grand principle. How different the Cosmogony of BUFFON—a man to whom genius was never wanting, who brightened every subject on which he touched, and kindled every mind which approached him, because, although with faculties imperfectly balanced, he lived and had his being amid the richest and rarest qualities of the philosophic spirit,—how different and wholly fantastic his idea that planets were chips struck off the sun by the collision of comets! Not one of the fundamental conditions of our system's mechanism could be explained by this wild and reckless imagination, whereas Laplace's bold and brilliant *induction* (may I not now so name it?) includes and resolves all! The theory is so beautiful and so perfect, that perhaps we might have assumed it to be universal, and asserted that every planet springing out of rotation, and engirdling each of those infinite orbs must be subjected to the chief laws which control the earth — had not presumption been checked by one emphatic indication. If, as we are informed, the two least problematical Satellites of Uranus have retrograde motions, *i. e.* if they move in directions opposite to the ge-

neral one, there must be some influence or law
capable of checking in so far and modifying
the operations indicated by the Nebular Cosmogony; and this intimation reaches us from
the farthest verge of our system—that confine
where novel external actions would be the most
sensibly felt.   Whatever this influence is, it
cannot invalidate the theory of Laplace.   The
laws of nature never destroy but only modify
each other,—just as the systems of circular waves
diffused from two centres in a pool, intermingle
and affect each other's undulations—each spreading meanwhile out to the extreme limits of the
sphere.

We have been advancing very rapidly—
let us pause now and look back on the grand
perspective below us.   We set out by asking,
can the Nebular Hypothesis *explain the stars?*
Somewhat indeed remains to be fathomed, and
phenomena apparently disparate may still be
found in the sky; yet, short way as we have

gone, every one of the grand features of the
stars—facts, which but to mark, have often
worthily conferred deathless fame—are seen in
union and harmony the most unexpected, proceeding hand in hand from the bosom of previous night, and going through untold ages in
singular companionship,—as if we had even attained the privilege of witnessing the arranging
influence of that Dove Spirit, which erst brooded
over chaos. It is indeed an impressive spectacle!
Who can ascend so far up that vast chain which
unites the eternal past with the fleeting present; who—to go no higher—can dwell on the
idea of our Sun being born from one of those
dim nebulæ, order growing within him by effect
of law, and the worlds he illumines and sustains, springing gradually into being,—without
engrossing emotions! Sometimes on contem-
plating this mighty progression, and thinking of
the changes, visible and concealed, which must
have marked the advance of an organization so
majestic; asking, too, what is Man, save a

transient organization, with whose progress the
education of a Spiritual Being has been for
a moment connected—I confess I have been
so fanciful as to doubt whether those great and
good men who endowed the stars with spiritual
principles, ought to be deemed mistaken,—
whether that orb, during its fathomless evolu-
tions, may not have been the seat of a SPIRITUAL
POTENTATE, gifted with the glorious capacity
to rise in knowledge, power, and beneficence,
by experience of all the vast events of which
he is the centre—whether we should not look
upon these HOSTS of HEAVEN, as something
still more awful than inanimate worlds fitted to
sustain a Life like ours?—Far as our ken has
reached between us and the HIGHEST there is
still vastness and mystery :—sometimes to take
wing beyond terrestrial precincts, perhaps, is not
wholly forbidden ; provided we go with unsan-
dalled feet, as if on Holy ground.

Apart however from all speculation,—surely
the view of an actual order whose beginnings

are hid in what seems in our eye nothing less
than Eternity, cannot but elevate our thoughts
of that BEING, who amid change alone is un-
changeable—whose glance reaches from the
beginning to the end—and whose presence
occupies all things! If uneasy feelings are
suggested—and I have heard of such—by the
idea of a process which may appear to substitute *progress* for *creation,* and place *law* in the
room of *providence,* their origin lies in the mis-
conception of a name. LAW of itself is no
substantive or independent power; no causal
influence sprung of blind necessity, which car-
ries on events of its own will and energizes
without command. Separated from connexion
with an ARRANGER in reference to whose mind
alone, and as expressive of the Creative Idea it
can be connected with the notion of control—
Law is a mere name for a long order—an order
unoriginated, unupheld, unsubstantial, whose
floor sounds hollow beneath the tread, and whose
spaces are all void; an order hanging tremblingly

over nothingness, and of which every constituent
—every thing and creature fails not to beseech
incessantly for a substance and substratum in
the idea of ONE—WHO LIVETH FOR EVER !

IV. I have fully detailed to you all the physical adaptations of the Nebular Hypothesis ;
but in its application to the planetary system
it possesses besides, a peculiar philosophic ex-
cellence, or an accordance so exquisite with all
similar causes in Nature, that I cannot leave it
without further remark. It chiefly rests in
this department, on its power to explain the
system's first or constituent elements ; now
these remarkable features exercise most important influences over the system's destiny.
If you have rightly apprehended the principle
on which the stability of a planet's orbit depends
—the balance, viz. of its centrifugal force,
and the power of the sun's attraction—you
will perceive that the influence of any third

body upon the planet might disturb this permanence. But the idea of gravity had not
long been present to Newton's mind, before he
saw that as it is a universal power—a power
diffused from every particle of matter within
the system ;—not only must the sun attract the
planets, but *each planet must attract every
other ;* so that upon each individual body, there
is playing, as well as the sun's attraction, the
combined and ever-varying attractions of its
shifting companions. It appeared accordingly,
to our admirable countryman, that the path of
no planet could be perfectly stable, but on the
contrary, that every orbit must be constantly
moved somewhat from its place by these un-
steady influences; and he thought moreover,
that the disturbances thus introduced would
abate the symmetry of existing arrangements,
and in all likelihood ensure their destruction
by one irregular shock. The dissipation of
Newton's fear was reserved for that age whose
terminations we are still touching; when the

subtle analysis of the illustrious LAGRANGE,
instructed us that every one of the disturbances
referred to must be PERIODICAL or oscillatory,
i. e. if the moon is now approaching the earth,
or her orbit drawing in—a time must come
when that approach will cease, and when an
opposite action, or a retrograding will take
place, which also will be stopped in subsequent ages by a superior limit : and so of
all other perturbations.   This truth results
from the existence of those very constituent
elements of our system on which the Nebular Hypothesis is founded ; for if the planets
and satellites had not moved in the same direction alike in their revolutions and rotations,
and if the spheres within which they move had
not approached to circles, and been comprised
almost in the plane of the sun's equator—*this
periodicity would not have supervened, and no
antidote been found to Newton's sorrowful fore-
bodings.*   It were pleasing here to arrest our progress and admire so signal an illustration of that

beauty of design and beneficence of purpose visible in every portion of the great chain of causes and effects; but passing the attractive theme we rather hasten to conclusions directly in view. Those constituent elements are thus the *elements of the strength* of planetary arrangements ; and it is in closest connexion with *them* that we have found the originating cause of the whole planetary system. Now, in every department of organized life—whether vegetable or animal—it is a law that in the progress of reproduction and growth, we have an especial adaptation to produce or ensure the existence of such *essential* circumstances, while the causes of the accidental ones are evanescent and of irregular action; and perhaps it is its fine accordance with this wide and striking analogy, its adaptations to that portion of planetary phenomena which, more than all others, must have sprung directly from the causes of our system's birth, which have gained for the theory of Laplace the respect of all, and the silent acquiescence of many Astronomers.

But farther,—the system, though strong, is not framed to be EVERLASTING; and our Hypothesis also developes the mode of the certain decay and final dissolution of its arrangements. Remember the effects of the Solar Ether! Although no mark of age has yet been recognised in the planetary paths, as sure as that filmy comet is drawing in its orbit, must they too approach the sun, and at the destined term of their separate existence, be resumed into his mass.* The first indefinite germs of this great organization, provision for its long existence, and finally its shroud, are thus all involved in that master

---

* It may be asked does not this Ether rotate along with the planets and therefore not retard them? It must rotate somehow—the comets will one day discover that for us; but it can-not rotate with velocities corresponding to all the planets. Nay, the very ellipticity of the planetary orbits, small as this is, necessitates a retardation in every one of them, however the Ether may rotate.

conception from which we can now survey the
mechanisms amid which we are ! And mark the
nature of this decay.   It comes, not as Newton
thought, by accident, derangement, or disease,
but through the midst of harmony; it is an easy
consequence of the venerable power which first
evolved us, infused our scheme with the spirit
of life, and gave it structure and strength. Our
supposed origin of the planets gave them and
their satellites that kind of orbits, and that kind
of rotation which produced their permanence ;
and the inherence of this same Nebulous
parentage, viz., the existence of an ether, leads
gently to their decline. So dies Nature's un-
blemished child—the simple flower! It bursts
its seed, buds and blooms; and then in un-
pained obedience draws in its leaves and sinks
into the lap of its Mother Earth.

The idea of the ultimate dissolution of the
solar system, has usually been felt as painful,
and forcibly resisted by philosophers.   When

Newton saw no end to the deranging effect
of the common planetary perturbations, he
called for the special interference of the Almighty to avert the catastrophe; and great
was the rejoicing when that recent Analyist, descried a memorable power of conservation in our
system's constituent phenomena; but after all,
why should it be painful? Absolute permanence
is visible nowhere around us, and the fact of
change, merely intimates, that in the exhaust-
less womb of the future, unevolved wonders are
in store.   The phenomenon referred to would
simply point to the close of one mighty cycle in
the history of the solar orb—the passing away
of arrangements which have fulfilled their objects that they might be transformed into new.
Thus is the periodic death of a plant perhaps
the essential to its prolonged life, and when
the individual dies and disappears, fresh and
vigorous forms spring from the elements which
composed it.   Mark the Chrysalis!   It is the
grave of the worm, but the cradle of the sun-

born insect. The broken bowl will yet be healed and beautified by the potter, and a Voice of joyful note will awaken, one day, even the silence of the Urn!

—Nay, what though *all* should pass? What though the close of this epoch in the history of the solar orb, should be accompanied, as some by a strange fondness have imagined, by the dissolution and disappearing of all these shining spheres? Then would our Universe not have failed in its functions, but only been gathered up and rolled away these functions being complete. That gorgeous material framework, wherewith the Eternal hath adorned and varied the abysses of space, is only an instrument by which the myriads of spirits borne upon its orbs, may be told of their origin, and educated for more exalted being; and a time may come, when the veil can be drawn aside—when spirit shall converse *directly* with spirit, and the creature gaze without hinderance on the efful-

gent face of its Creator : but even, then—no,
not in that manhood or full maturity of being,
will our fretted vault be forgotten or its pure
inhabitants permitted to drop away. Their
reality may have passed, but their remembrance
will live for ever. The warm relationships of de-
pendant childhood, are only the tenderer and the
more hallowed, that the grave has enclosed and
embalmed their objects; and no height of excel-
lence, no extent of future greatness, will ever
obscure the vividness of that frail but loved infancy, in which, as now, we walked upon the
beauteous earth, and fondly gazed upon these
far-off orbs—deeming that they whisper from
their bright abodes the welcome tidings of Man's
immortal destiny !

LETTER VIII.

SPECULATION.

IF our desire of knowledge did not quicken as its sphere expands, or if man, so long as an eminence is unsurmounted, could lay himself down in peace, satisfied with the view of the vastness and variety which already stretch out beneath him, doubtless our task had now ended, and the volume of Astronomy might have been closed. But Desire, happily insatiable, has no confine on this side the Infinite ; and no sooner have we reached the elevation of one thought or idea which resolves some large portion of the unknown, than ambition is fired afresh, and speculation never at rest, takes wing towards remoter regions.

In the present instance indeed there is every

N

encouragement to further venturous inquiry.
The Nebular Hypothesis, in its relations to the
planetary system, may be termed complete;—
it comprehends its beginnings, establishes those
elements on which its duration depends, and
exhibits the causes and mode of its ultimate
transition into a novel form; and thus—surveying it from its commencement to its close—we
are as if in possession of that primeval Creative
Thought which originated our system and planned and circumscribed its destiny. Now, in
reference to one of these epochs, our Hypothesis seems to hold equal connexion with the
whole contents of the Heavens,—the epoch,
viz. of their origin; and is it—as a conclusion
of our task—too daring to fancy that likewise
in *their* progression under the control of law
from primitive chaos, principles are evolved
which regulate the existing distribution of stars
into firmaments, which have determined the
form of the present condition of Universe, and
from which new forms will issue unceasingly,

until as with the planets the hour of final transmutation comes, and the cycle of their existence is complete? In endeavouring to pursue such an inquiry—supported, as we are only by a few very remote hints from observation—there is need of all our cautiousness; nor can we hope to obtain farther than the most general idea of what DEITY keeps in store for those majestic stellar arrangements.

I. There is something very remarkable in the aspect of the more simple firmaments, as affecting the question of their probable permanence. In the first place, it is not difficult to conceive mechanisms by which a simple and regular globular or oval cluster would be retained in absolute and everlasting stability. We have indeed only to suppose that every sun in it describes an oval or ellipsis around the general centre of attraction, and *all in the same time.* In whatever direction the bodies might move, whether direct or retrograde, in the same plane

or across each other's orbits, such a cluster
would be permanent; and during all time its
orbs would thus revolve, arriving together after
some vast interval at the exact position from
which they departed,—which interval would be
the great year, the *annus magnus* of the cluster.
Even although no such precise motions should
any where exist, this fact is extremely valuable.
It illustrates the possibility of the existence of
motions among the orbs of these brilliant masses,
by whose relationships the influence of gravity
might be permanently withstood; and it per-
mits the conception, that whether or not that
influence is *absolutely* withstood there, still they
may be subject to adjustments not less efficient
and beautiful than the constituent arrrangements
of our own system, by which harmony is pre-
served in the face of change, and strength
during a long decline.

Such a condition of permanent stability,
however, depends on the absence of all resisting
causes. Should retarding causes exist, the recur-

rence of the Annus Magnus would no longer find
the different orbs in their first position, but to a
certain amount nearer the centre of attraction,
and the whole cluster somewhat compressed :
so that if a nebulous fluid—the remnant of
ancient chaos—still pervades the intervals of
these orbs, compression must ensue, and the firmament would pass through all degrees of it,
occupying ages in its mysterious growth towards some other form of being.   Parting from
the notion of a cluster of perfect regularity, the
idea thus rises before us, of a *series*, in which
various compressions will be exhibited, and upon
which MUTABILITY is stamped.   Now, *this
series exists ;* if we would characterise the globular clusters we have resolved, and without reference to theory, it would be in language we
have already used (see Letter V.), and by their
various compressions.   These objects present
a series quite unbroken ; and they are in the
exact condition illustrated by Herschel when
he compared them with plants in different stages

of progress, from adolescence to proximate ripening (which in this case is decay)—precisely as Laplace afterwards most aptly characterised the varying aspects of the Nebulæ. It is not possible that these phenomena can be mere illusions. A *real* series in Nature is never meaningless; and where it seems so, we have assuredly committed the error of assuming that to be a series between whose different facts or stages there is no connexion—where we have no evidence of an evolving or transmuting cause.

Simple as these considerations are, another large field for contemplation is now opened. Even the Heavens are not stable! These globular masses at least, are in process of growth, of *ripening* — they are congregating together towards that nucleus, around which the new order of things is slowly up-growing, and where the mighty orb, foretold by their aggregating, is preparing to be born. I cannot avoid reverting to the notion of Mr Cole-

ridge ;—what is this after all, save a prolonga-
tion of the condensing of a Nebula ?   Already
some few of its particles have come together
and formed its secondary stage, and now that
secondary stage, which we term a firmament,
is passing into a third, where all the dispersed
atoms will be gathered together, and lodged
at the centre of the mass !

Fancied *relations* between the different clusters have often been the theme of speculation.
Probably they are related ; nay, it is almost
certain they are so : in which case that new
order we are supposing would resolve itself into
mighty orbs, separated by corresponding intervals, and related as the small stars which
now exist.   We would then, in fact, have but
one vast cluster, in which what are now firmaments would be units, a cluster which, if the
Nebulous matter as it seems to do, pervades all
space, must be subject to the same laws as before, and pass towards a similar termination.

The eye is wearied in stretching out towards the times thus foreshadowed,—they are epochs with which the duration of all that we conceive is utterly incommensurable; but yet their coming is sure, and by the existing forms and arrangements of matter their features are all preparing.

II. On turning from these simpler arrangements, to that which assuredly is most complex, but which one day we will know best— our own Firmament, aspects become distinguishable not only sustaining the foregoing conclusions through a strong analogy, but pointing the way to still bolder thoughts. The milky way has been already described as a ring for the most part isolated, in which the stars are very dense, and where the aggregating power has drawn them from the general mass, and by some curious operation, compressed them into a crowded girdle. But neither is this girdle uniform. In Letter V. I called your especial

attention to its division, apparently into groups
for the most part inclining to the spherical form,
and separated from each other by dark spaces
like wrinkles of age. Sir William Herschel
counted no less than 225 such groups or subordinate clusters, within the extent of it he examined; and as all these were of a kind to mark
the action of gravity, he concluded the existence
of a clustering power, drawing the stars of it into
separate groups, a power which had broken up the
uniformity of the Zone, and to whose irresistible
power it was still exposed. " Hence," says Herschel in one of those bold moments in which he
fearlessly traversed the infinities alike of past and
future, " Hence may we be certain that the stars
will there be gradually compressed through successive stages of accumulation, till they come
up to what may be called the ripening period of
the globular cluster and total insulation; from
which it is evident that the milky way, must
forcibly be broken up, and cease to be a stratum of scattered stars." " We may also," he

continues, in the same lofty mood, " draw an important additional conclusion from the gradual dissolution of the milky way ; for the state into which the incessant action of the clustering power has brought it, is a kind of *chronometer*, that may be used to measure the time of its past and present existence; and although we do not know the rate and going of this mysterious chronometer, it is nevertheless certain, that since a breaking up of the parts of the milky way, affords a proof that it cannot last for ever, it equally bears witness that its past duration cannot be admitted to be infinite !" Surely the vision of these unfathomable changes—of the solemn march of the majestic Heavens from phase to phase, obediently fulfilling their awful destiny, will be lost on the heart of the adorer, unless when beneath the canopy on which their annals are inscribed, it swells with that humility which is the best homage to the Supreme !

Grounding on these sublime speculations, and

taking along with us some other facts, we may perhaps safely rise still farther. If the aggregation of the stars in the milky way still go on as it prognosticates for Ages, the clusters, now with some intermission forming its ring, will become isolated, and appear in the character of separate systems. But, if this may happen in time future, *may something similar not have happened in time past ?* Referring to that approximate chart in Plate II., how irregular it is, how narrow in one direction, and how ragged its edges ! Can it be possible that masses of stars have been torn away from it in that direction, so that its thinness may simply indicate, that through the action of some irresistible cause, parts of it had there ripened sooner ? Singular to relate, it is precisely towards these thin sides, and almost immediately beyond them, that the vast mass of *neighbouring* isolated clusters is found — clusters all spherical, and grouping together in extraordinary proximity. In Plate XXII., you have part of the wing of VIRGO—

a constellation situated at right angles to the milky way, and accordingly at the shallowest part of our firmament. Observe how crowded it is with groups—some of them Nebulæ indeed, but most of them small round clusters, exhibiting a marked degree of compression. In regions on the opposite side, the same phenomenon recurs; from which, aided by speculation, we infer that the process of separation has gone on, and that the breaking up of the milky way in our time, may simply be a repetition of part of the changes of the long past, to which our capricious firmament owes its irregular form.

Nay, are not even these operations only types of what may have occurred on a far more majestic scale? The separate firmaments which our telescope has descried and mapped down, show, even more emphatically than the groups in the milky way, the efficacy and progress of a clustering power;— may not *they* have come originally from one

Plate XXII.

Forrester & Nichol lithog.

homogeneous stratum or mass of stars, so that their present isolation — their separation and various grouping, are only the ongoings of the clock—the gigantic steps of the hand, by which Time records the days of the years of the existing mechanism of the Universe! Stupendous the conception, that these great Heavens—the Heavens we have deemed a synonym of the Infinite and Eternal, are nothing other after all than one aspect in which matter is destined to present itself—their history like the birth, life, the death and dissolution of the fragile plant! If this indeed is true, and on behalf of the conception we can marshal many probabilities, how immense the sphere of real existence—how little can we ever know of it; at least, how much must be referred to that higher state of existence—an expected eternity of sublime contemplation.

But this is not all :—Take also into view the relations of this vast mechanism to the moral world; observe the amount of most diverse organization, the almost infinite varieties of intelligence—wonderfully various, even within our limited ken—which must be attached to the progress of the Heavens, growing with them as they grow, changing during every new phase, and sympathizing with their decline :—who then shall say that even the highest created spirit will ever exhaust the fulness of that volume which God has spread before us all, in illustration of his own Infinite Nature !

Suppose we are yet mistaken; suppose the Nebular Hypothesis with all its comprehensiveness not to be the true key to the mystery of the origin and destinies of things, what is gained—what new possession—by that course of bold conjecture on which we have ventured to embark ? This at least, is established on grounds not to be removed. In the vast Heavens, as well as among phenomena around us, all things are in a state of change and PRO-GRESS ; there too—on the sky—in splendid hieroglyphics the truth is inscribed, that the grandest forms of present Being are only GERMS swelling and bursting with a life to come ! And if the universal fabric is thus fixed and constituted, can we imagine that aught which it contains is unupheld by the same preserving law, that annihilation is a possibility real or virtual—the stoppage of the career of any advancing Being, while hospitable Infinitude remains ? No ! let the night fall ; it prepares a dawn, when Man's weariness will have ceased, and his soul be refreshed and restored.— To COME !—To every Creature these are words of Hope spoken in organ tone ; our hearts suggest them, and the stars repeat them, and through the Infinite, Aspiration wings its way, rejoicingly as an eagle following the sun.

Farewell!

NOTES.

NOTE A.

THE SOLAR SYSTEM.

THE object and scope of my treatise did not admit of the incorporation with it even of a brief exposition of the phenomena of the solar system,—our contemplations inclining us to regard the sun with his attendants as one unit, and presupposing, in some instances, a general acquaintance with the system. That no doubt may rest, however, over our speculations, because of the absence of a view of the system's details, I have thrown into this note an account of most of what is important on the subject.

1. The solar system is composed of a majestic central luminary (whose mass is made up of matter like the earth—the atmosphere alone being luminous), and a number of small engirdling bodies, which revolve around him in various periods. Toys named Orreries have generally been used to give an idea of this mechanism :— they are mere toys, often beautiful, but never instructive; nor indeed can they be otherwise, as you will learn from the statement of the distances, and comparative magni-

tudes of the several bodies, which I extract from Sir John
Herschel's work :—" Choose any well levelled field or
bowling green.   On it place a globe two feet diameter ;
this will represent the SUN ; MERCURY will be represented by a grain of mustard seed, on the circumference
of a circle 164 feet in diameter from its orbit ; VENUS, a
pea on a circle 284 feet in diameter ; the EARTH, also a
pea on a circle of 430 feet ; MARS, a rather large pin's
head, on a circle of 654 feet ; JUNO, CERES, VESTA, and
PALLAS, grains of sand in orbits of from 1000 to 1200
feet ; JUPITER, a moderate sized orange, in a circle nearly
half a mile across ; SATURN, a small orange on a circle
of four-fifths of a mile ; and URANUS, a full sized cherry
or small plum, upon the circumference of a circle more
than a mile and a half in diameter."   Such are the contents and relative dimensions of the solar system !

2. After the labour and observation and mistakes of
ages, it became a settled point with instructed astronomers, that these small bodies, including the earth, move
around the sun, which is the proper centre of the system ;
and the question then started, in what curves or paths do
the planets revolve ?   A decisive answer was first ela-
borated by the illustrious JOHN KEPLER, who not only
proved that they move in ellipses, or ovals, differing little
from circles, but also established a law for the varying
velocity of each planet when in different parts of its
orbit, and the existence of a great relation between the
distance and velocity of one planet, and the distance and

velocity of any other. The three statements expressive of these general facts, are termed Kepler's Laws, and were our first accurate step towards a knowledge of the true mechanism of the Heavens.

3. After Kepler NEWTON arose, and completed the superstructure founded by his great predecessor. His first achievement was the combining of Kepler's three principles into one, and the constitution of dyanimical astronomy. It is known as a universal fact, that bodies when set in motion tend to move in straight lines, and would so move for ever with uniform speed, provided no external force or influence interposed to deflect them from their natural paths. For instance if the body A were set in motion in the direction A B, it would move of it-

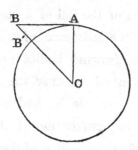

self in the straight line A B; and if we find it instead, moving in the curve A B′, the inference is, that an external influence has deflected it, or turned it from its natural course; the effect of which influence in the case in the diagram, clearly is to have drawn the body from B to B′ in the time it would have occupied in moving from A to B. On the ground of these principles Newton re-

solved the orbitual motions of the planets, by referring
them to two forces—their natural tendency to move in
straight lines through space, and an attractive or gravitat-
ing force drawing them to the centre of the sun; and by a
most refined and novel method of demonstration, he
proved that the attractive force exercised by the sun
over the *different* planets, is precisely the *same power*
diminishing in intensity in a certain manner according
as the distance of the planet increases.   How simple did
the celestial mechanism thus become!   Each planet pre-
serves its place and rolls through its appointed orbit,
in consequence of the balanced action of two simple and
regular forces!

4. But Newton did not stop here.   Meditating on
the every-day phenomenon of the fall of a stone to the
earth, it struck him that the existence of a force was
here indicated, efficient in drawing bodies towards the
earth ; and he daringly asked whether that force diminished according to the principle he had already detected, would not explain the gravitation of the Moon?
The discovery lay in the first suggestion of the idea, for
the pursuit of it was not difficult.   If, for instance, A B
in the foregoing diagram, is the space the moon would
move in a straight line in one second of time if left to
her first natural impulse, and A B' her actual orbit, the
quantity B B', which is quite susceptible of measurement, will, manifestly show the power of the earth's attraction over her in one second: and this, when placed

beside the quantity through which a stone at the earth's surface falls in the same time, exhibits two motions having to each other the exact proportion required by the celestial law of gravity!   It is impossible to overestimate the value of this momentous step.   It immediately *connected the celestial mechanism with terrestrial phenomena,* establishing a first analogy between the Heavens and the Earth, and yielding an emphatic intimation of the great Unity of Things.   Newton now generalized his first conception ; announced gravity as an universal property of matter—a law obeyed by every particle in reference to every other particle ; and thus threw open the gates of the august Temple of the Universe.

Subsequent astronomers have merely applied his comprehensive principle to the solution of cosmical phenomena.   The orbits of comets became at once intelligible.   The figures of the planets were explained ; and the determination of those fine influences, by which the planets affect each other, has constituted the chief employment of dyniamical science until the present day. It is one object of my little work to collect and arrange what traces have been found of the action of the same fundamental law among the external Stellar spaces.

NOTE B.

THE ORBITS OF THE DOUBLE STARS.

THE curious and inquisitive will naturally desire to obtain some notion of the actual *size* of the orbits of these double systems,—or how far one star is distant from the other. Were we acquainted with the distance of a binary system from the Earth or Sun, it would be quite easy to calculate its dimensions ; but until that previous knowledge is obtained, all information derived from this source must be purely conjectural. In the *Annuaire du Bureau des Longitudes,* for 1834, M. ARAGO has with his usual precision detailed a very ingenious method, first pointed out by SAVARY, which promises to afford accurate results on the subject. It depends upon the fact that Light does not move instantaneously, but with a certain definite velocity, so that a specific time elapses between the moment when the ray leaves a luminous body and that when it enters an eye. The following

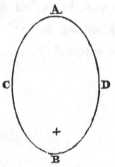

considerations will illustrate the method I speak of. Suppose A C B D the orbit of a double star, whose period is 200 years, and let the star be so distant from the earth when at the points A and B of its orbit, that its light occupies two years in the first instance and one year in the second, in passing toward us,—what will be the effect of this upon the body's apparent motion? The result will be brought out most distinctly by calculation.

Let the star be at A in the year 1800,

It will be at B in . . 1900,

And at A on its return in 2000.

But since light occupies two years in arriving from A to the Earth, and one year in reaching us from B,

The star will be seen by us at A in 1802,

And at B in . . . 1901.

So that it will seem to have performed one half of its revolution, or to have gone through the half A C B of its orbit in *ninety-nine* years.

Again,

It is seen at B in . 1901,

And once more at A in 2002.

So that it will seem to have revolved from B to A, or to have gone through the other half of its orbit B D A in *one hundred and one* years. But as it really does pass through these two arcs in precisely the same time, this apparent difference of its periods must be owing to the *size of its orbit as measured by the velocity of light;* and in fact half the difference of the two numbers 101 and

99, or one year, *is just the period occupied by light in traversing the distance from A to B.* The rule is quite general : observe the apparent times occupied by any revolving star, in going through the two halves of its orbit ; and half the difference of these times will be the period in which light passes through the diameter of its orbit : and as the velocity of light is known, that diameter may therefore be computed in miles. The only obstacle to the general application of the method consists in the difficulty of noting exactly when the star is at the two opposite points of its orbit ; but notwith-standing of this difficulty it will yield results accurate within certain limits. It is impossible to give too much credit to the ingenuity of M. SAVARY.

NOTE C.

RELATION OF THE NEBULAR HYPOTHESIS TO GEOLOGICAL

CHANGE.

I MEAN to compress into this Note a few remarks at one time destined for the text, regarding the influence which may be exercised by the Nebular Hypothesis over the course of geological speculation.

It does not require notice that our Hypothesis supplies the matter of the planets, &c. in a condition sufficiently plastic to permit of its taking on the precise form in each case, which a solid of revolution ought to have ; nor indeed is this of any considerable moment, inasmuch as a great number of hypothetical cosmogonies offer similar facilities. When we go farther however, and trace the subsequent physical history of the bodies thus fashioned, we discover peculiar adaptations of our theory to the solution of problems which pure geology has not hitherto comprehended, and in regard of which that science must sooner or later borrow aid from Astronomy. The causes of geological change are two—an *abrading* and *levelling cause,* operating chiefly on the surface of our globe through the agency of running water; and an *elevating cause*—that which has upheaved our mountain masses, and which is still efficient not only in partial volcanic actions, but over large tracts of territory, such as Sweden, which are slowly but gradually rising to a higher elevation. The difficulty in geology is to discover the nature of this elevating cause, and it is with this question, as well as the mode of its operation in time past, that the conflict of geological theories is now concerned.

We enter on the subject most easily and naturally through Dr Daubeny's theory of Volcanoes. It is well known that the metallic *bases* of the earths, alkalis, &c. inflame with considerable deflagration on contact with water; and this philosopher supposes that masses of these unoxidated metals lie beneath the surface of the earth in subterranean reservoirs, to which water from the surface may find occasional access, and cause an explosion. Chemical objections have been alleged against this theory, but they are of no cogency; and I believe that existing volcanic phenomena receive from it the best and most logical explanation which has yet been proposed. Let us now express this theory in general language having a view to the Nebular Hypothesis. Our world is in that epoch of its being which, physically speaking, may be termed the *epoch of the fluidity of water,* and we find that this water, which is the nebulous bed or stratum last condensed, exerts chemical actions upon the previously condensed bed of so violent a nature,

that deflagration and explosion issue from their fortuitous meeting. How easily is this theory extended! In
its very nature it points to a general law; for what is
going on with regard to these two most recent beds, may
have supervened through all time past with any two strata
which came in contact: the stratum B, in every stage,

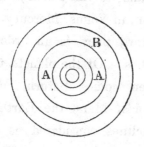

may have had chemical relations with the stratum A, so
that *a deflagrating and elevating power may have characterised every epoch of the history of every planet.*

When Daubeny's theory has received this general
expression, the knot is cut which so much puzzles geolo-
gists of the present day, regarding the comparative extent
of modern and ancient volcanic forces. Mr Lyell has
achieved a great philosophical reform in this department
of inquiry, in boldly asserting it as a fundamental tenet
that we must philosophize upon the ground of existing
and known forces; but when he passes beyond this and
denies the possibility of the superior energy of these
forces in former epochs, he oversteps his limits, and
dogmatizes,—because it is only when the nature of the

origin of the force is comprehended, that we can ascertain whether or not it is plastic. The Nebular Hypothesis, as above provides for its variability; and existing astronomical phenomena seem to establish that it *is* variable.

It has long been known that the elevating cause—whatever it is, is not peculiar to this Earth;—its operations are seen in the Moon, in Venus, Mercury, and perhaps also in the Sun. Now, there is an order of age among these celestial bodies. Of the planets, the eldest is the farthest off, then comes Saturn, afterwards Jupiter, the four Asteroids, Mars, the Earth, Venus, Mercury; and the Satellites considered as complete worlds, are perhaps the most recent of all. The inspection of the different planets then, shews us this elevating cause in operation in a considerable variety of epochs, and we find, that the more advanced or the older the planet, the less rugged or mountainous is it—the less energetic this shattering power. Saturn and Jupiter show no appearance of mountains; those of Venus, and in all probability of Mercury, are gigantic, when compared with the existing elevations of the Earth; and when we reach our own Moon, we find a picture of that torn, crateriform, and disturbed surface, which our own earth may have exhibited when the masses of existing mountains were thrown up, when the pinnacles of the oldest ranges were unabraded, and the ruggedness of the valleys not yet smoothed away by the levelling power of the fluid which deposits and stratifies there, the detritus of the primary rocks.

It is much to be wished that when a science arrives at its ultimate problems, we would no longer study it in isolated fashion, but look abroad through the Universe, for those relations which constitute it a part of the grand whole.

NOTE D.

I subjoin a statement of the exact places of the more important objects represented in the foregoing plates ; so that the possessors of good telescopes may have the opportunity of examining such as are within their reach. I give their Right Ascensions and Declinations, by means of which their localities among the stars may easily be detected upon a celestial globe, and thus found without difficulty in the Heavens.

	Right Ascen.		Declinations.	
	H.	M.	Deg.	Min.
Plate I. Cluster in Hercules,	16	35½	36	45 N.
Plate III. .	13	22½	46	14 N.
Plate V. Fig. 1, . .	15	10	2	44 N.
Fig. 2, . .	21	25	1	34 S.
Plate VI. Fig. 1 (counting from the left top of the line and passing across the page) 30 Doradus,	5	39½	69	15 S.
Fig. 2, .	0	16½	73	0 S.
Fig. 3, .	21	31	23	55 S.
Fig. 4, .	15	29	6	33 N.
Plate VII. .	4	55	72	

	Right Ascen.		Declinations.	
	H.	M.	Deg.	Min.
Plate VIII. Fig. 1, faintest of the				
two rings,	20	9½	30	3 N.
Fig. 2, brightest of two				
rings, .	18	47	32	49 N.
Fig. 3, oblong hoop,	2	12	41	35 N.
Fig. 4, oval figure be-				
side the hoop,	12	48	22	37 N.
Fig. 5, large figure	19	52	22	16 N.
Plate XII. Nebula in Orion,	5	27	5	30½
Plate XIII. Fig. 1, Nebula in				
Andromeda, .	0	33½	40	20
Plate XIV.				
Fig. 2, .	20	38½	30	6 N.
Fig. 3, .	20	49	31	3 N.
Plate XV. . .	18	11	16	15 S.
Plate XVI. Double Nebulæ,	9	22½	22	15 N.
	7	15	29	49 N.
	22	51	13	43 S.
Two stars on faint bed of light	18	7	19	56 S.
Plate XIX. Nebulous stars,	17	0	23	7 N.
	19	40	50	6 N.
	3	58½	30	20 N.
Plate XX. Reticulated form,	20	50	29	34 N.
Nebulous matter con-				
nected with stars,	12	51	35	47 N.
	20	56	11	24 N.
	6	30	8	53 N.
	8	46½	54	25 N.

P

ADDITIONS AND CORRECTIONS.

Page 49, line 5, *for* Plate II., *read* Plate I.

Pages 35 and 36—It should not be inferred from the account given of the forty-feet telescope, that we are much indebted to it for actual discovery. The truth is, it was so unwieldy, mechanically speaking, that comparatively little use could be made of it. In the text I refer only to its optical powers.

Page 72, line 3 from the bottom, *for* ξ, *read* ζ.

Page 89, line 4, *for* 88, *read* 80.

Page 114, line 9, *for* " first of our learned societies," *read* " same learned society."

Page 132, line 12, *for* " characteristics or peculiarities," *read* " characteristic peculiarities."

Page 179.—At line 3 from the bottom, I refer to two of the Satellites of Uranus. Only these two have been seen by any observer since the time of Sir William Herschel. It is scarce to be doubted that their motions are retrograde, and in orbits highly inclined to the ecliptic.

Printed in the United States
By Bookmasters